Wireless Sensor Networks

Wireless Sensor Networks

Evolutionary Algorithms for Optimizing Performance

Damodar Reddy Edla
Mahesh Chowdary Kongara
Amruta Lipare
Venkatanareshbabu Kuppili
Kannadasan K

CRC Press
Taylor & Francis Group
Boca Raton London New York

CRC Press is an imprint of the
Taylor & Francis Group, an **informa** business

A CHAPMAN & HALL BOOK

First edition published 2021
by CRC Press
6000 Broken Sound Parkway NW, Suite 300, Boca Raton, FL 33487-2742

and by CRC Press
4 Park Square, Milton Park, Abingdon, Oxon OX14 4RN

Typeset in CMR
by Nova Techset Private Limited, Bengaluru & Chennai, India

**Visit the Taylor & Francis Web site at
http://www.taylorandfrancis.com**

**and the CRC Press Web site at
http://www.crcpress.com**

ISBN: 978-0-367-61315-0 (pbk)
ISBN: 978-0-367-34241-8 (hbk)
ISBN: 978-0-429-32461-1 (ebk)

Contents

Preface

Wireless Sensor Networks (WSNs) is one of the emerging technologies over the last few decades. It has a number of applications in areas such as field monitoring, health care, surveillance, military applications, home applications and so many others. Along with the usage of the WSN, a number of risks and challenges occur while deploying any WSN. The sensor nodes are mostly battery-powered; therefore, energy-efficiency is crucial in WSN. This book focuses on achieving the prolonged lifetime of the WSN and proposes the energy-efficient algorithms for the same. Six bio-inspired algorithms are discussed in the book and the way of applying them to the WSN is revealed here.

In this book, Chapter 1, the introduction, shows challenges in the wireless networks. Chapter 2 discusses the basic background needed for understanding the remaining chapters. Chapter 3 through Chapter 8 discusses different evolutionary approaches for improving the performance of the wireless networks. A brief discussion of those chapters are provided below.

The description of our contributions in six technical chapters of this book is summarized below:

1. **Chapter 3: Shuffled Complex Evolution (SCE) Approach for Load Balancing:**

   In this chapter, Shuffled Complex Evolution algorithm is used for load balancing of gateways in WSN. A novel fitness function is also designed to evaluate fitness of solutions produced by SCE algorithm. In SCE, the solutions with best and worst fitness value exchange their information to produce new offspring. We have simulated proposed load balancing algorithms along with other state-of the-art load balancing algorithms, namely Node Local Density Load Balancing (NLDLB), Score Based Load Balancing (SBLB), Simple Genetic Algorithm (SGA) based load balancing, and Novel Genetic Algorithm (NGA) based load balancing.

2. **Chapter 4: Improved Shuffled Complex Evolution (ISCE) Approach for Energy Efficiency:** In this chapter, the improvements to the SCE are conducted in terms of various phases. The initial population phase, the crossover phase, has been modified to generate valid offsprings. A new phase is also added to SCE to improve the quality of the solution. A novel fitness function is designed in terms of distance ratio and the load ratio. The proposed ISCE is compared with state-of-the-art algorithms to validate its performance under various evaluation factors.

3. **Chapter 5: Improved Shuffled Frog Leaping Algorithm Approach for Load Balancing:**

   In this chapter, Shuffled Frog Leaping Algorithm (SFLA) is improved by suitably modifying the frog's population generation and offspring generation phases in SFLA and by introducing a transfer phase. A novel fitness function is also designed to evaluate the quality of the solutions produced by the Improved SFLA (ISFLA). We performed extensive simulations of the proposed load balancing algorithm in terms of various performance parameters. The experimental results are encouraging and demonstrated the efficiency of the proposed algorithm.

4. **Chapter 6: SCE-PSO Based Clustering Technique for Load Balancing in WSN**

   In this chapter, we proposed (1) a clustering algorithm based on the shuffled complex evolution of particle swarm optimization (SCE-PSO); (2) a novel fitness function by considering mean cluster distance, gateways' load and number of heavily loaded gateways in the network. The experimental results are compared with other state-of-the-art load balancing approaches, such as Score Based Load Balancing (SBLB), Node Local Density Load Balancing (NLDLB), Simple Genetic Algorithm (SGA) based load balancing and Novel Genetic Algorithm (NGA) based load balancing.

5. **Chapter 7: PSO Based Routing with Novel Fitness Function for Improving Lifetime of WSN:**

   Particle Swarm Optimization (PSO) based routing is proposed in this chapter. Also, a novel fitness function is designed by considering the number of relay nodes, the distance between the gateway to sink and relay load factor of the network. The proposed algorithm is validated under different scenarios. The experimental results show that the proposed PSO based routing algorithm prolonged WSN's lifetime when compared to other bio-inspired approaches.

6. **Chapter 8: M-Curves Path Planning Model for Mobile Anchor Node and Localization:**

   In this chapter, we propose a novel path planning approach for mobile anchor based localization called M-Curves. Our proposed model promises that all the nodes in the network will receive at least three non-collinear beacon messages for localization. Our proposed trajectory assures full coverage, high localization accuracy as compared to other static models. Also, we optimize the localization process by using Dolphin Swarm Algorithm (DSA). The fitness function used for optimization in DSA minimizes the localization error of the node in the network.

Authors

Dr. Damodar Reddy Edla is an Assistant Professor in the Department of Computer Science and Engineering at National Institute of Technology Goa, India. He received M.Sc. Degree from the University of Hyderabad in 2006, M.Tech. in Computer Application and PhD Degree in Computer Science and Engineering from Indian School of Mines Dhanbad, in 2009 and 2013, respectively. His research interests include Cognitive Neuroscience, Data Mining, Wireless Sensor Networks and Brain-Computer Interface. He has published more than 100 research articles in reputable journals and international conferences. He is a senior member of IEEE and IACSIT. He is also an Editorial Board member of several international journals.

Mahesh Chowdary Kongara received the B.Tech. Degree in Computer Science and Engineering from Sree Vidyanikethan Engineering College, Tirupati, Andhra Pradesh and the M.Tech. Degree in Department of Computer Science and Engineering from National Institute of Technology Goa, India, in 2017. He is currently a full-time Research Scholar with the Department of Computer Science and Engineering, National Institute of Technology, Goa, India. His research interests include Soft Computing, Wireless Sensor Networks and Internet of Things.

Amruta Lipare has received the B.Tech. Degree in Information Technology from Rajarambapu Institute of Technology, Sakharale, Maharashtra, in 2015, and the M.Tech. Degree in Computer Science and Engineering from National Institute of Technology, Goa, in 2017. She is currently a full-time Research Scholar with the Department of Computer Science and Engineering, National Institute of Technology Goa, India. Her research interests include Soft Computing, Wireless Sensor Networks, Evolutionary Computations and Swarm Intelligence.

Dr. Venkatanareshbabu Kuppili, PhD (IIT Delhi), is with the Machine Learning Group, Department of CSE, NIT Goa, India, where he is currently an Assistant Professor. He was with Evalueserve Pvt. Ltd., as a Senior Research Associate. He is also actively involved in teaching and research development for the Graduate Program in Computer Science and Engineering Department at the NIT Goa, India. He has authored several research papers published in reputable international journals and conferences. He is a senior member of IEEE.

Kannadasan K has completed his B.Tech in Information Technology at SASTRA University, Tamilnadu and M.Tech in Computer Science and Engineering, National Institute of Technology, Goa. Currently, he is pursuing his PhD degree at the National Institute of Technology, Tiruchirappalli. He is a student member in IEEE and Secretary at IEEE Student branch, NIT Tiruchirappalli, India. His research interests include Machine Learning, Wireless Sensor Networks, Swarm optimization techniques, Brain-Computer Interface, etc.

Chapter 1

Introduction

1.1 Introduction

Wireless Sensor Networks (WSNs) are networks that consist of a large number of small and low-energy sensor nodes. Such nodes are randomly or manually scattered across a given target area or territory. The sensor node contains the data aggregation unit, the sensing unit, the communication portion, as well as the energy unit. The sensor nodes in the network may have a Global Position System (GPS) to progress within the target area or territory. WSNs have promising applications in a variety of areas such as disaster warning systems, health care, environmental monitoring and protection and in crucial areas such as security, intruder detection and defence identification [1]. Sensor networks are used to collect data from the environment and build conclusions about the monitored object. Limited communication capabilities usually distinguish these sensor nodes due to power and bandwidth constraints. Therefore, reducing the energy conservation of sensor nodes in the network is a challenging task for WSNs. Energy-efficient clustering and routing algorithms are further studied in this regard [2], [3]. Figure 1.1 represents the architecture of a WSN. The basic components of the WSN are the following.

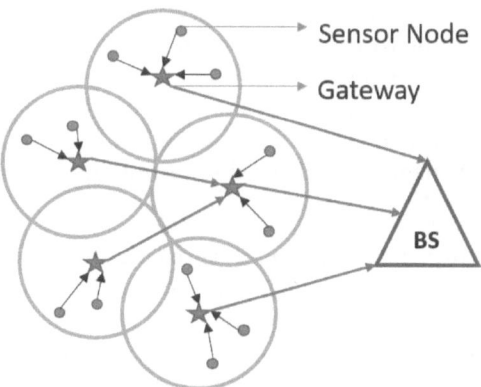

FIGURE 1.1: Architecture of WSN

1. **Sensor Node:** The sensor node is a basic unit in WSN and consists of a sensor unit, a data processing or aggregation unit, a communication unit and a power unit. The sensor is a component and type of transducer in the sensor node. The transducer is a device that converts one form of energy to another. The sensor node aims to detect changes in its environment and then provide the corresponding output. The sensor nodes consist of a microcontroller, a transceiver, a power source, an external memory and some sensors [4].

2. **Cluster Head/Gateway:** Sensor nodes are divided into several groups called clusters in WSN. Every cluster would have a leader named the Cluster Head (CH). The purpose of the CH is to collect local information from the sensor nodes in the group, aggregate the data and send it to the sink or base station. In some WSN cases, some of the sensor nodes in the network behave as CHs. In some WSN scenarios, high-level energy sensor nodes are deployed, called gateways, because the workload of CHs is massive compared to non-CHs [5].

3. **Sink/Base Station:** Sink or Base Station (BS) is a basic unit or component in WSNs. Sensor nodes collect local information in their environment. The CHs collect these data from sensor nodes and transmit them to the base station. These base stations connect to the internet [6].

1.2 Challenges in WSNs

The following issues influence the wireless sensor networks design [7, 8, 9, 10].

1. **Energy Efficiency:** Energy efficiency involves less energy to provide the same service or to do the same amount of work. Energy efficiency is an essential design factor in WSNs, and it mainly depends on the energy consumption of nodes in the network. The proper usage of energy in the system can increase the network lifetime of WSN. Clustering and routing are two effective optimization techniques to improve the lifetime of WSN.

2. **Load Balancing:** Load balancing is one of the most critical factors in the design of WSNs. Load balancing means assigning sensor nodes to CHs in such a way to maximize the existence of the network. The load can be considered as the number of packets transmitted by the sensor nodes or gateways. It can also be considered as the number of neighbour nodes of the sensor nodes, or the nodes in the communication range of the gateway.

3. **Fault Tolerance:** Sensor nodes signal strength, antenna angle and obstacles, and weather conditions trigger WSN fault tolerance. For example, the sender sends the data to the receiver, but the receiver is unable to receive the data. In this scenario, the sender utilizes energy for data transfer, but there is no improvement in network operation.

4. **Localization:** Localization can be termed as computing the position of the sensor node deployed in the network. GPS can be used for localizing the sensor node. Due to limited power resource of sensor nodes, high cost and poor performance of GPS in the indoor environment, GPS is not an efficient solution for localization. Therefore, developing an algorithm for localization is one of the challenges in WSN.

1.3 Motivation

Wireless sensor networks offer massive benefits for several industries such as building automation, energy management, substation control, industrial monitoring and structural health monitoring. Sensor nodes can be installed in cities to monitor and control the concentration of dangerous gasses and detection of fire in the forest. Due to the lack of battery power, limited storage space, communication and computational capacities, the design of efficient algorithms is essential for improving the lifetime of the network. Clustering sensor nodes and routing for data transfer are two strategies that can be used to achieve the goal of optimizing network lifetime. In cluster-based WSN, some CHs may get overloaded due to improper cluster creation. This congestion causes a great deal of damage to the network, such as increased communication latency, high CH energy consumption and its shows overall performance of the WSN. Thus, the load balancing of the CHs is the most vital concern for clustering sensor nodes. During routing, some CHs may have excessive energy usage when sending data to the base station or sink due to improper data routing. Also, the deployment of sensor nodes with GPS is very costly. Therefore, to make the deployment cost-efficient, we proposed the localization approach using the anchor nodes.

1.4 Objectives and contributions of the book

The contributions presented in this book shall attempt to address the following objectives using bio-inspired algorithms for various challenges.

- Load balancing of gateways using Shuffled Complex Evolution (SCE) algorithm.

- Improving the SCE algorithm in various phases to generate an energy-efficient solution.

- Designing a novel fitness function for a bio-inspired algorithm called Shuffled Frog Leaping (SFLA) used for energy efficiency in WSN.

- Developing a clustering algorithm using a Shuffled Complex Evolution of Particle Swarm Optimization called SCE-PSO.

- Generating an optimal solution for routing of data packets using PSO.

- Designing an algorithm for localization of sensor nodes using Dolphin Swarm Optimization Algorithm.

To meet the above objectives, the description of our contributions in six chapters of this book is summarized below:

1. In **Chapter 3**, we have applied the SCE algorithm to balance the load of the gateways in WSN. SCE is modified according to the application of WSN. The solution is represented as the assignment of sensor nodes and gateways in WSN. In the initial population phase, instead of assigning nodes randomly, we assigned sensor nodes to gateways in their communication range. The novel fitness function is designed to calculate the quality of the solution. The proposed algorithm is compared with various state-of-the-art algorithms to validate the performance of the proposed algorithm.

2. In **Chapter 4**, we have applied the SCE algorithm with various improvements for energy efficiency and load balancing of the gateways in WSN. The modifications are made in the various phases of SCE. A new phase is added to obtain the optimal solution from the population. The novel fitness function is designed according to the distance to the base station and energy of the sensor nodes. The proposed ISCE algorithm is compared with some of the existing algorithms under various evaluation factors.

3. SFLA is an evolutionary algorithm used for various complex optimization problems. In **Chapter 5**, we have applied SFLA for energy efficiency and load balancing of gateways in WSN. The SFLA is also modified according to the application of WSN. Like SCE, The solution is represented as the assignment of sensor nodes and gateways in WSN. In the initial population phase, instead of assigning nodes randomly, sensor nodes are assigned to gateways in their communication range. The novel fitness function is designed to calculate the quality of the solution. The partition of the population into memeplexes and sub-memeplexes is carried out according to the factors of the size of the population. The proposed SFLA-based algorithm is validated with some of the existing algorithms.

4. In **Chapter 6**, we have adopted SCE-PSO algorithm for a load-balanced problem with a novel fitness function. In clustered-WSNs, the load of the gateways and energy usage of sensor nodes are the most crucial parameters. The trade-off between the two objectives is an essential factor for enhancing the lifetime of WSNs. Therefore, the novel fitness acknowledges a load of gateways as well as mean cluster distance between sensor nodes and gateways. The fitness value is used to measure the efficiency of the particles. The SCE-PSO helps to reach an optimal solution fast.

5. In **Chapter 7**, we have proposed PSO based routing for sending data to the sink and also a novel fitness function to acknowledge the quality of the generated solutions. The novel fitness considers the maximum distance between the gateway and the sink, hop-count and relay load factor of the network. The trade-off between three objectives can improve the routing path of the network. The weighted sum approach can help to choose the importance value of the purposes. The relay load factor can benefit from choosing a better routing path for data transfer.

6. In **Chapter 8**, we propose a novel path planning approach for mobile anchor-based localization called "M-Curves". Our proposed model promises that all the nodes in the network receive at least three non-collinear beacon messages for localization. Our proposed trajectory assures full coverage, high localization accuracy as compared to other static models. Also, we optimize the localization process by using Dolphin Swarm Algorithm (DSA). The fitness function used for optimization in DSA minimizes the localization error of the node in the network.

1.5 Resources used

This research work is conducted with the help of different resources provided by our institute. The institute has provided some of the e-resources such as IEEE Xplore, ScienceDirect, ACM Digital Library and Springer Link, etc. These resources were beneficial for continuing this research work. We have also referred to different online journals such as *IEEE Transactions on Wireless Communications, Applied Soft Computing, Computers and Electrical Engineering, Swarm and Evolutionary Computation, Engineering Applications of Artificial Intelligence, Wireless Networks, Wireless Personal Communications*, etc. We also referred to the proceedings of several international conferences which were very valuable for carrying out our research.

We also refereed various books on WSN, such as *Fundamentals of Wireless Sensor Networks: Theory and Practice* [66], *Wireless Sensor Networks: A Networking Perspective* [67], *Wireless Sensor Networks: From Theory to Applications* [68], *Wireless Sensor Networks: Architectures and Protocols* [69],

Building Wireless Sensor Networks: With ZigBee, XBee, Arduino, and Processing [70], *Deploying Wireless Sensor Networks: Theory and Practice* [71], *Soft Computing in Wireless Sensors Networks* [72], *Wireless Sensor Networks and Applications* [73], *Innovations in Swarm Intelligence* [74], *Swarm Intelligence and Bio-inspired Computation: Theory and Applications* [75], *Evolutionary Algorithms in Engineering and Computer Science: Recent Advances in Genetic Algorithms, Evolution Strategies, Evolutionary Programming, GE* [76], *Hybrid Evolutionary Algorithms* [77], *Multi-objective Optimization Using Evolutionary Algorithms* [78], *Evolutionary Algorithms in Engineering Applications* [79], *Bio-inspired Algorithms for Engineering* [80] and *Bio-Inspired Computational Algorithms and Their Applications* [81].

The various web pages, specifically from Google and Google Scholar, were very supportive of acquiring useful information. The comments and suggestions received from several reviewers and editors were beneficial and facilitated in improving the quality of our research work. For the experimentation, we used MATLAB R2015a on an Intel i7 Processor machine with 3.60 GHz CPU and 8 GB RAM running on Microsoft Windows 10 Pro, which is provided by our department. We used Microsoft Office and LaTeX for documentation and presentation purposes.

1.6   Organization of the book

The rest of the book is structured as follows:

Chapter 2 reviews a brief survey on different load balancing algorithms, clustering, routing, energy efficiency and localization algorithms used for WSN.

Chapter 3 introduces an algorithm based on SCE for load balancing of gateways in WSN. It also discusses an overview of SCE and presents the comparative analysis between various state-of-the-art approaches and the proposed approach.

Chapter 4 proposes various improvements in SCE called ISCE, applied for energy efficiency as well as load balancing of gateways in WSN. It also shows the comparative study of ISCE against various existing algorithms.

Chapter 5 proposes an SFLA-based algorithm for energy efficiency in WSN. It also discusses the overview of SFLA. Moreover, the performance of the proposed approach is discussed.

Chapter 6 presents a clustering approach using SCE-PSO algorithm. The preliminaries related to this are also discussed here. The proposed approach is validated with various existing algorithms.

Chapter 7 explains an algorithm for routing of data packets using the PSO approach. The PSO algorithm is also discussed in this chapter. The performance of the proposed approach is explored in the results.

Chapter 8 proposes a localization approach for WSN. The accuracy of the positions of the sensor node is calculated. The optimal solutions are generated by applying Dolphin Swarm Optimization.

Chapter 9 discusses the conclusion of all the chapters and possible directions for future research work.

1.7   Conclusion

This chapter provided an overview of the book. The motivation behind this research, objectives and scope is described. In the next chapter, we provide a comprehensive survey related to the research work.

Chapter 2

Literature Survey

In this chapter, we discuss various existing techniques for load balancing, energy efficiency in WSN as well as the localization problem in WSN. Firstly, we focus on an extensive review of various challenges in wireless sensor networks and their heuristic approaches. We then discuss different meta-heuristic approaches and the way they are applied to various problems in WSN. Another challenging issue in WSN is localization. Therefore, we discuss localization-related works further in Section 2.3.

2.1   Heuristic approaches

In the literature, various algorithms regarding CH selection and cluster formation are described. One of the famous algorithms among them is Low-Energy Adaptive Clustering Hierarchy (LEACH) algorithm [11]. In LEACH, the CHs are selected based on predefined probability. Each sensor node gets a chance of working as CH after each P number of rounds. Once the CH is selected, other sensor nodes are connected to their nearest CH forming the new clusters at each round. Member sensor nodes send data to the CH according to the schedule provided by CH. After a certain number of rounds of transmission, all the sensor nodes drain their energy, still taking part in the CH competition. So, at that round, no transmission is carried out. To overcome this drawback, Kumar et al. [12] came up with the improved LEACH (I-LEACH) algorithm. In I-LEACH, the CHs are selected according to the probability as well as the lifetime of the sensor nodes, i.e., remaining energy of sensor node.

Jana et al. [13] have proposed a load-balancing algorithm using a heap data structure. The sensor nodes are categorized into two types: i) open sensor nodes and ii) restricted sensor nodes. The sensor nodes in the communication range of one and only one gateway are known as restricted sensor nodes. The sensor nodes in the communication range of two or more than two gateways are known as open sensor nodes. In the algorithm, all of the restricted sensor nodes are allocated to their gateways in range. Later on, the assignment of open nodes is carried out according to the min-heap data structure. The parameter for building the min-heap is considered as the distance of the sensor node to the gateways.

Zhang and Yang [14] have proposed a load-balanced clustering algorithm by considering the local density of each gateway. The local density of node is nothing but the number of neighbour nodes of that sensor node. In this algorithm, initially, the sensor nodes in range (R) of one and only one gateway are allocated to the respective gateways. In the second step, the sensor nodes which are within the range of R/2 are connected. In the last step, the sensor nodes are connected to the gateways in range and with least node density.

In [15], the clusters are formed first, and then CHs are selected from the cluster according to the residual energy. The score of each sensor node is calculated by the ratio of residual energy and distance of sensor node from the BS. The sensor node with the maximum score is selected as the CH for the next round.

In [16], Baranidharan and Santhi defined a new fuzzy approach for load balancing in WSN. The fuzzy approach is distributed in two phases: (i) cluster formation phase and (ii) data collection phase. The fuzzy inference system evaluates each sensor nodes according to its residual energy, distance to the BS and node density. The CHs are elected according to its chance and residual energy. The clusters are formed according to the size of the sensor node.

Avoiding energy hole in the network is one of the challenges in WSN. So, to reduce energy hole Prasenjit et al. [17] have proposed a balanced load scheme. The CHs are selected according to the on-demand selection of CH system. A load-balanced routing technique is used to save energy consumption in the network.

Morteza et al. [18] defined another approach to reduce the energy hole in the network. The clustering is carried out at each level. The CHs are selected at the next level according to the residual energy and the distance of a node to the BS. The sensor nodes near to the BS directly send data to the BS as well as they route the data from the sensor nodes far from the BS.

2.2   Meta-heuristic approaches

Along with heuristic approaches, various meta-heuristic approaches for energy efficiency in WSN are also available in the literature.

For load balancing of gateways, Kuila et al. [19] have proposed a novel genetic algorithm (NGA) based approach. The genetic algorithm [82] is an evolutionary algorithm which generates random solutions and improves the solution at each generation. In NGA, the solution is represented as the set of sensor nodes and their assignment to the gateways. To evaluate the solution, the authors have proposed a novel fitness function.

Al-Aboody et al. [20] have applied a bio-inspired algorithm for energy efficiency in WSN called Grey Wolf Optimizer (GWO). GWO [83] is a meta-heuristic algorithm inspired by grey wolves. Authors have used GWO for selection of CHs and a tree-based approach for routing of data packets. Here, in the fitness function, only the residual energy of sensor node is used and tried to avoid the formation of an energy hole at the BS.

One of the popular bio-inspired algorithmic techniques is Ant Colony Optimization (ACO) [21, 22, 23] inspired by the behaviour of ant colonies and this work postulate. Many researchers have applied ACO as an application to the mobile sink. Ants have very less capability of exploring the local area, but they are good at achieving global performance. One of the examples of global performance is finding the minimum distance path from the location of ant to the food source. Ants discharge a fluid called pheromones to mark their path while traversing their journey. The probability of the specific path chosen by an ant is proportional to the concentration of the pheromone, where pheromone concentration differs via evaporation or reinforcement. With this mechanism, finally, ants find their food source. Song and Cheng-lin [84] have applied ACO for unequal clustering in WSN.

The GA-ABC [24] is a hybrid algorithm which uses Genetic Algorithms (GA) and Artificial Bees Colony (ABC) [85] for the cluster formation in WSN. ABC is a swarming algorithm inspired by the bees. GA-ABC is defined into two stages: i) Setup phase and ii) Data aggregation phase. GA is used to select the appropriate CHs in among the sensor nodes, whereas ABC helps to join the other sensor nodes to the CHs. In the phase of data aggregation, the sensed data from the sensor nodes are collected at the CHs and, in turn, the CHs send data to the other CHs towards the BS. At last all the data is collected at the BS.

In 2006, the T-ANT [25] was introduced as a distributed routing protocol. T-ANT stands for two-phase clustering process using ACO. The two phases are the cluster formation phase and the communication phase. In the cluster formation phase, the CHs are selected by applying the ACO, and BS implements the series of ants called control messages. The communication is conducted with the help of these control messages. The main aim of T-ANT is to distribute energy and minimize energy consumption uniformly.

The Particle Swarm Optimization - Centralized (PSO-C) [26] is one of the bio-inspired protocol based algorithms used to find CHs. PSO-C works similar to the LEACH-C [27] algorithm. It minimizes the intra-cluster distance between the CHs and cluster member nodes and reduces the power consumption in the network.

2.3 Localization-related work

Various localization models and mobility models for mobile anchor-based localization have been proposed. Mobility models can be classified into four categories [42]. The first category, Static Anchor Static Node, where both the anchor nodes and sensor nodes are static. This type of model is proposed in [43, 44, 45]. The second category, Static Anchor Mobile, is where the anchor node is static while sensor nodes move inside the deployed environment[46, 47]. The third category, Mobile Anchor Static Node, which we are interested in, is where the anchor node moves around the network and sends beacon signals, whereas the sensor nodes are static in the environment. The fourth category, Mobile Anchor Mobile Node, is where both the anchor node and sensor nodes move around the network [48, 49].

Further, the movement strategy of the mobile anchor can be classified into static path mobility and dynamic path mobility. The difference between static path mobility from other models is that the trajectory is fixed and cannot be changed or modified once the nodes are deployed. Hence, most of the static paths are designed based on trilateration or triangulation. Generally, static models show high localization ratios as these models have as one of their main goals assuring that all the nodes receive the beacon messages for localization. However, points to be considered while proposing a path are features of anchor points and issues with the path like collinearity and path length.

The two major trajectories proposed for mobile anchor-based localization are SCAN and HILBERT [38]. In SCAN, mobile anchor moves in one dimension in straight lines. *Resolution* is defined as the distance between any pair of consecutive lines. Also, the distance between two anchor points in the path is represented as resolution. SCAN model is the simplest model in terms of implementation, but its accuracy is less because of the collinearity problem.

The trajectory followed by the SCAN model is shown in Figure 2.1 (a). HILBERT was proposed to overcome the issues of SCAN with turns in the trajectory. Deployment area is divided into four squares of equal sizes, and the points are connected to make the trajectory. Path model of HILBERT is shown in Figure 2.1 (b). HILBERT gives better performance compared to SCAN as it gives three different beacon positions. HILBERT had a coverage problem as the trajectory did not go through the borders of the network. Unknown sensor nodes deployed along the border of the network did not receive enough beacon messages to localize themselves thus resulting in high localization error.

Based on trilateration technique, a trajectory for the mobile anchor is proposed in [40] named as LMAT. The path is designed to overcome the collinearity issue by following the shape of symmetric triangles such that each unknown node receives three different non-collinear beacon messages. In LMAT, resolution is defined as the distance between every two anchor points. Though it achieved high localization accuracy, the issue with LMAT is its path length. The trajectory of LMAT is shown in Figure 2.1 (c).

A superior path planning model termed as Z-Curves is proposed in [39]. The mobile anchor follows the shape of 'Z' in its trajectory. The network is divided into squares for three levels. Level 1 follows the 'Z' pattern, and in the successive levels, the 'Z' patterns in the previous level are connected as shown in Figure 2.1 (d). As like the other path models, Z-Curves focus on

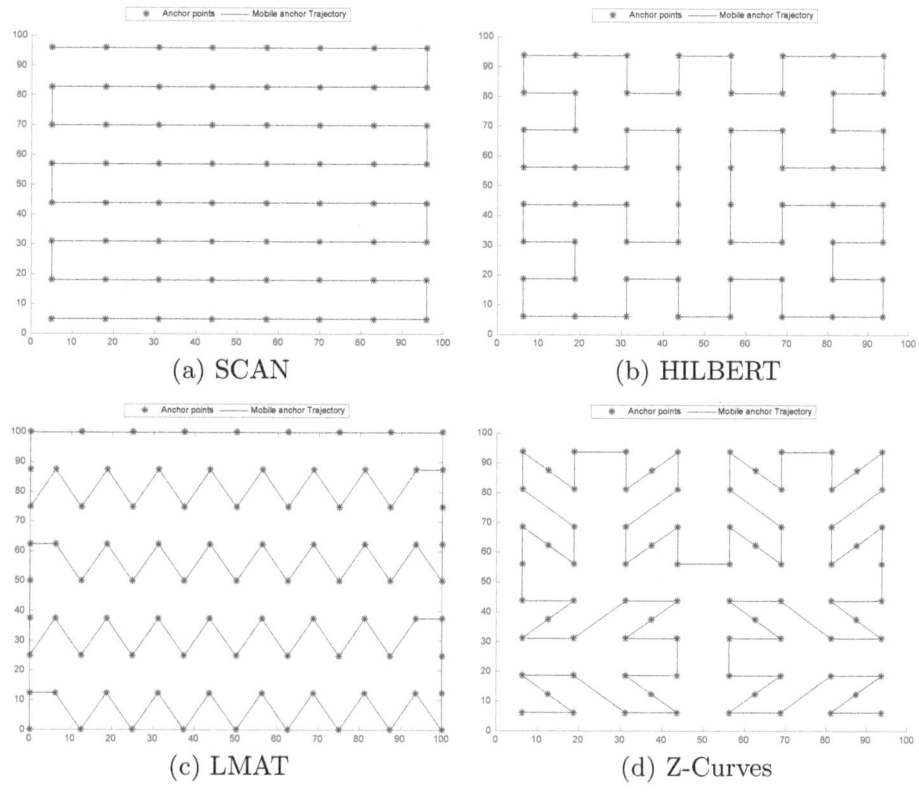

(a) SCAN (b) HILBERT

(c) LMAT (d) Z-Curves

FIGURE 2.1: Mobile anchor trajectory for various static models

the trilateration concepts and ensure that the unknown nodes receive three different beacon messages.

2.4 Conclusion

In this chapter, we presented a comprehensive review of several existing heuristics and meta-heuristic approaches used for defeating various challenges in WSN such as load balancing, energy efficiency, clustering, routing and localization. In the next chapter, we develop a shuffled complex evolution-based approach for load balancing of gateways in WSN and investigate its performance.

Chapter 3

Load Balancing of Gateways Using Shuffled Complex Evolution Algorithm

3.1 Introduction

Shuffled Complex Evolution [58] is an evolutionary algorithm used for local and global search optimization. The SCE-UA method is a general-purpose global optimization program. It was originally developed by Dr. Qingyun Duan as part of his doctoral dissertation work at the Department of Hydrology and Water Resources, University of Arizona, Tucson, AZ 85721, USA. The dissertation is entitled "A Global Optimization Strategy for Efficient and Effective Calibration of Hydrologic Models". The program has since been modified to make it easier for use on problems of users' interests.

In this chapter, we discuss how to use the SCE-UA method [58] efficiently and effectively for load balancing of gateways in a wireless sensor network. The competent local search feature of the SCE algorithm is suitable for solving different optimization problems. One of the challenging issues in WSN is load balancing of gateways. The solution of the load balancing is considered as the assignment of the sensor nodes and gateways.

3.2 Preliminaries

3.2.1 Energy model

We have used the same energy model as used in [86]. This radio model for energy consists of both free space and multi-path channels according to the distance from the transmitter to the receiver. So, the energy required for transmission of l-bit data over distance d is given by Equation 3.1.

$$E_T(l,d) = \begin{cases} l * E_{elec} + l * \epsilon_{fs} * d^2, & d < d_0 \\ l * E_{elec} + l * \epsilon_{mp} * d^4, & d \geq d_0 \end{cases} \tag{3.1}$$

Where d_0 is the threshold, if the distance is more than threshold value then multi-path fading model has been used; otherwise free space model is used. E_{elec} is the energy required by electronic circuitry, ε_{fs} is energy required by the free space and ε_{mp} is the energy required by multi-path channel. The energy required to receive l-bit of data is given by Equation 3.2.

$$E_R(l) = l * E_{elec} \tag{3.2}$$

The E_{elec} includes some features like the digital coding, modulation, filtering and signal distribution. The distance from the transmitter determines the amplifier energy, $\varepsilon_{fs} * d^2$ or $\varepsilon_{mp} * d^4$, to the receiver as well as the acceptable bit error rate.

3.2.2   An overview of shuffled complex evolution algorithm

Shuffled Complex Evolution (SCE) [58] is an evolutionary algorithm performing a local and global search. A local search is done through a memetic evolution to evaluate a solution. Global search is carried out by changing the information from parallel local searches [59, 60]. SCE algorithm is discussed below.

Step 1: **Generating initial solutions**
Let X be the set of s number of points (solutions) $x_1, ..., x_s$ in its realistic space $\Omega \in R^n$. Calculate fitness value at each point x_i.

Step 2: **Ranking of points**
Sort points in decreasing order and store in an array D. $D = \{x_i, f_i | i = 1, ..., s\}$. The value at i=1 gives the points with highest fitness value.

Step 3: **Partitioning the solutions**
Partition the array D in p complexes such that $A_1, A_2, ..., A_p$. Every complex contains m points in such a way that the first point is allocated to the first complex, the second allocated to the second complex, point p is allocated to the pth complex, and point (p + 1) is allocated to the first complex, and so on.

Step 4: **Evaluating each complex**
Each complex A_k is evaluated according to competitive complex evolution (CCE) algorithm. CCE algorithm is described further.

Step 5: **Shuffling the complexes**
Substitute all complexes into array D. Again sort D with decreasing order according to their fitness value.

Step 6: **Checking convergence**

Check whether convergence criteria is satisfied or not. If it has been satisfied stop, else go to step 3.

Competitive Complex Evolution (CCE) Algorithm: The steps in CCE algorithm [61] are described below.

Step 1: **Initializing variables:** Choose q, α, β, where $2 \leq q \leq m, \alpha \geq 1$, $\beta \geq 1$.

Step 2: **Assigning weights:** The triangular probability distribution has been assigned to A_k, as follows:
$p_i = 2(m + 1 - i)/m(m + 1)$, where $i = 1, \ldots, m$.

p_1 will have the maximum probability because $p_1 = 2/m + 1$. Point p_m will have minimum probability because $p_m = 2/m(m + 1)$.

Step 3: **Selection of parents:** In this selection process, q distinct points $u_1, \ldots, u_k$ are randomly chosen from A_k according to distribution of probability described in *Step 2*. These q points are outlined as a sub-complex and stored in array B, where $B = \{u_i, v_i | i = 1, \ldots, q\}$, and v_k is fitness value associated with point u_k. The locations of A_k are stored in L and are used to build B.

Step 4: **Generation of offspring:**

(a) Arrange t points in decreasing order of their fitness value in order to sort B and L. Centroid g is computed by using the following expression.
$g = [1/t - 1)] \sum_{j=1}^{t-1} u_j$

(b) Reflection step: Calculate new point $r = 2_g - u_q$.

(c) If new point is better than worst point, then replace worst point. Else, a contraction point is calculated. Contraction point is at midway between new point and worst point.

(d) If contraction point is better than the worst point then substitute to worst point. Else, within the feasible domain, a random point has generated and the worst point is replaced by this random point.

(e) The steps b to d are repeated for α times, where $\alpha \geq 1$.

Step 5: **Combining offspring into single population:** A_k has been sorted in decreasing order and stored in B; also their locations are stored in L.

Step 6: **Iteration:** Repeat *Steps 2* to *5* for β times where $\beta \geq 1$.

3.3 Proposed load balancing algorithm

In this section, the proposed load balancing algorithm using SCE optimization is discussed. In the following subsections, we explain the critical steps from SCE algorithms, namely representation of individuals, initial population generation, fitness function evaluation, parent selection and the technique of partitioning into complexes and sub-complexes.

3.3.1 Individual representation

An individual is represented as the mapping of sensor nodes and gateways. We have considered the length of the individual as the number of sensor nodes. Example 1 shows the concept of individual representation.

Example 1: Consider a WSN of 4 gateways and 12 sensor nodes, i.e., $G = \{G_1, G_2, G_3, G_4\}$ and S= $\{S_1, S_2, \ldots, S_{12}\}$. As number of sensor nodes are 12 therefore the length of individual for this network is 12. Table 3.1 shows individual representation for this WSN. 'S' is the set of sensor nodes and 'G' is the set of gateways. From Table 3.1 it is interpreted as S_4, S_5 and S_6 are allocated to G_1, G_4 and G_3, respectively.

3.3.2 Initial population generation

The set of individuals (solutions) is called an initial population. Each individual is generated randomly with an assignment of sensor nodes and gateways. The solution is said to be valid if all the sensor nodes from the solution are in the communication range of the gateway assigned to the respective sensor node. A valid solution represents the complete clustering solution. This initial population generation is explained with an example given below:

Example 2: Consider a WSN of 4 gateways and 12 sensor nodes, i.e., G=$\{G_1, G_2, G_3, G_4\}$ and $S = \{S_1, S_2, \ldots, S_{12}\}$. As there are 12 sensors, the length of each individual is 12. Table 3.2 shows gateways in range of each sensor node. A sensor node can be allocated to any of the gateways in their range. According to Table 3.2, sensor node S_1 is assigned to either gateway G_1 or G_2 and so on.

TABLE 3.1: Individual representation

S	1	2	3	4	5	6	7	8	9	10	11	12
G	1	2	1	1	4	3	2	3	4	3	2	4

TABLE 3.2: Sensor nodes and their corresponding gateways within communication range

Sensor Nodes	Gateways in Range
S_1	G_1, G_2
S_2	G_2, G_3, G_4
S_3	G_1, G_4
S_4	G_2, G_3
S_5	G_1, G_4
S_6	G_1, G_2, G_4
S_7	G_2, G_4
S_8	G_2, G_3
S_9	G_1, G_2, G_3, G_4
S_{10}	G_1, G_3
S_{11}	G_1, G_2, G_3
S_{12}	G_1, G_3, G_4

TABLE 3.3: Initial population generation

S	1	2	3	4	5	6	7	8	9	10	11	12
G	2	4	1	2	4	4	4	3	1	1	2	3

From the example 2, sensor node S_3 has selected gateway G_1 among G_1 and G_4; likewise S_1 selects gateway G_2 among G_1, G_2, S_2 selects G_4 among G_2, G_3, G_4, and so on. This forms the initial population, which is shown in Table 3.3.

3.3.3 Proposed novel fitness function

We have designed an efficient fitness function to evaluate each individual solution given in Equation 3.3. For efficient load balancing, we have considered a metric called expected gateway load in the fitness function. This metric calculates the average gateway load on every gateway, and it can be computed using Equation 3.4.

$$Fitness = \text{Expected gateway load} * \frac{\text{Number of granted gateways}}{\text{Total number of gateways}} \quad (3.3)$$

$$\text{Expected gateway load} = \left[\frac{\sum\limits_{i=1}^{m} \text{Load Ratio } (G_i)}{\text{Number of gateways}} \right] \quad (3.4)$$

Load Ratio (G_i) term in Equation 3.4 is used to represent the overall load of gateways which is scaled from Load (G_i) term. It is computed using Equation 3.5.

$$\text{Load Ratio } (G_i) = \frac{\text{Load } (G_i)}{arg \max \text{Load } G_i, \forall i = 1, 2, \ldots m} \quad (3.5)$$

Load (G_i) represents the amount of energy consumed by gateway G_i in order to transmit the l number of bits. It is calculated using Equation 3.6.

$$\text{Load } (G_i) = l * \text{Percentage of remaining energy in } G_i \tag{3.6}$$

In order to identify the number of granted gateways, every gateway is checked for overloaded or underloaded gateway. To check the status of a gateway, two thresholds are defined: maximum load threshold and minimum load threshold. Maximum load threshold is used to check the overloaded status, and minimum load threshold is used to check underloaded status of gateway. These thresholds are defined using Equation 3.7 and Equation 3.8, respectively, where *Mean* is calculated using Equation 3.9 and γ is constant.

$$\text{Maximum threshold} = Mean + \gamma * Mean \tag{3.7}$$

$$\text{Minimum threshold} = Mean - \gamma * Mean \tag{3.8}$$

$$Mean = \frac{\sum_{i=1}^{m} \text{Load } (G_i)}{m} \tag{3.9}$$

From Equation 3.6, it is clear that the best solution is an energy-efficient solution due to the consideration of the remaining energy. In the solution, the number of sensor nodes to be assigned is proportional to the residual energy of the gateway. This helps in reducing the gateway energy consumption and maximizing the network lifetime. Hence, the proposed fitness function is proficient for load balancing as well as energy efficiency. Fitness function value varies from 0 to 1. The solution with fitness value 0 is the worst solution (entirely unbalanced network) whereas solution having fitness value 1 is the best solution (fully balanced network). Therefore, for better solution fitness, function needs to be maximized. The process of fitness value calculation for unbalanced and balanced networks is shown below by using Example 3 and Example 4, respectively.

Example 3: Consider there are four gateways G_1, G_2, G_3 and G_4 with loads as 30, 14, 20 and 8, respectively, and $\gamma = 0.25$.

From the above example it is clear that Load (G_1)= 30, Load (G_2)= 14, Load (G_3)= 20 and Load (G_4)= 8.
Compute the Load Ratio on every gateway using Equation 3.5, i.e.,
Load Ratio(G_1): $\frac{30}{30} = 1$, Load Ratio(G_2): $\frac{14}{30} = 0.47$, Load Ratio(G_3): $\frac{20}{30} = 0.67$ and Load Ratio(G_4): $\frac{8}{30} = 0.27$.
Expected gateway load is calculated using Equation 3.4,
i.e., $\frac{(1+0.47+0.67+0.27)}{4} = 0.6025$
Calculate the Mean using Equation 3.9, i.e., $\frac{(30+14+20+8)}{4} = 18$.

Then define Maximum threshold using Equation 3.7, i.e., $18 + (18*0.25)$ $=22.50$ and Minimum threshold using Equation 3.8, i.e., $18 - (18*0.25) =$ 13.50.

Load on G_1 is 30, which exceeds the maximum threshold limit of 22.50. Hence, G_1 is overloaded and not granted. Load on G_2 is 14, which is in the range of minimum and maximum threshold limit of 13.50 and 22.50. Hence, G_2 is granted. Similarly, G_3 is also granted, and G_4 is not granted as it is underloaded. Therefore, only G_2 and G_3 gateways are granted.

Finally, the fitness value is calculated using Equation 3.3, i.e., $0.6025 * \frac{2}{4} =$ 0.3012.

Example 4: Consider an ideal case where there are 5 gateways and the load on every gateway is 5 and $\gamma = 0.25$.

From the example, Load $(G_1)=$ Load $(G_2)=$ Load $(G_3)=$ Load $(G_4)=$ Load $(G_5)=5$. Load Ratio$(G_1)=$Load Ratio$(G_2)=$Load Ratio$(G_3)=$Load Ratio$(G_4)=$Load Ratio$(G_5)=1$.

So, the expected gateway load is 1, the minimum threshold limit is 0.75 and the maximum threshold limit is 1.25. Since all gateways are having a load of 1, which is in the range of minimum and maximum threshold limit, therefore, the number of granted gateways are 5. The fitness value is 1, which means the network is fully balanced. In real-time scenarios, it is not possible to obtain fitness value 1. For a better solution, we have to choose a solution that corresponds to the maximum fitness value between 0 and 1.

3.3.4   Sorting and partitioning of individuals

The fitness value of each individual is calculated using fitness function. All the individuals are then sorted in descending order (according to fitness value) and stored in an array D, where $D = \{x_i, f_i | i = 1, \ldots, s\}$. Here first element in array D represents a solution with highest fitness value. Array D is partitioned into p complexes. In this partitioning process, the first individual is allocated to first complex, second individual allocated to the second complex, p^{th} individual allocated to the pth complex, $(p + 1)^{th}$ individual is allocated to the first complex and so on.

This process has been explained for the Example 2. There are 12 individual solutions taken randomly as initial population. Let us consider set of individuals $I = \{i_1, i_2, \ldots, i_{12}\}$. These twelve individual solutions are evaluated and stored in an array D such that i_1 has highest fitness value and i_{12} has lowest fitness value. This is shown in Figure 3.1. Then an array D is partitioned into p complexes. This process is shown in Table 3.4 for $p = 3$.

In Table 3.4, A_1, A_2 and A_3 are three complexes. Now these complexes are divided into a number of sub-complexes (v). This process is shown in Table 3.5, for $v=2$, where sub-complexes will be denoted as (A_i, V_j) where i is complex number and j is sub-complex number.

Algorithm 1 Pseudo code for SCE-based Load Balancing

- **Input**
    - A set of sensor nodes $S - s_1, s_2, \ldots, s_n$.
    - A set of gateways $G = g_1, g_2, \ldots, g_m$, where $m < n$.
    - $Com(s_i)$: Set of all those gateways which are within the range of communication of s_i.
    - $PopSize =$ number of initial solutions.
    - A set of solutions $X = x_1, x_2, \ldots, x_{PopSize}$.
    - Xb= best solution in individual complex.
    - Xw=worst solution in individual complex.
    - Xg=global best solution among all complexes.
    - $P = \text{fact}(PopSize)$, i.e., $\text{fact}(PopSize)$ returns factors of $PopSize$.
    - $Q = \text{fact}(P)$.

- **Output**
    - An assignment A: S $\rightarrow$ G

- **Algorithm**
    1: **Step 1:**
    2: **for** $k = 1$ to β **do**
    3:     **for** $i = 1$ to n **do**
    4:         Assign sensor node s_i to their corresponding gateway g_i, randomly, such that $G \in Com(s_i)$.
    5:     **end for**
    6: **end for**
    7: **Step 2:**
    8: Apply fitness function for every solution in X.
    9: Sort X according to their fitness value.
    10: **Step 3:**
    11: Store X in array D with decreasing order.
    12: Choose a random number $p \in P$.
    13: Partition array D in p complexes, say $A_1, A_2, \ldots, A_p$, each containing n points, such that $D_1 \in A_1, D_2 \in A_2, \ldots, D_p$ in $A_p, D_{p+1} \in A_1$.
    14: Store D_1 as X_g.
    15: **Step 4:**
    16: **for** $i = 1$ to p **do**
    17:     Select v points randomly in A_i such that $i \le v < n$.
    18:     Make v sub-complexes from p complexes.
    19:     Select a point $q \in Q$.
    20:     Exchange information of all the sensor nodes after q point between Xb and Xw.
    21: **end for**
    22: **Step 5:**
    23: Evaluate each new offspring produced in **Step 4**.
    24: If convergence criteria are satisfied then Stop; else goto **Step 3**.

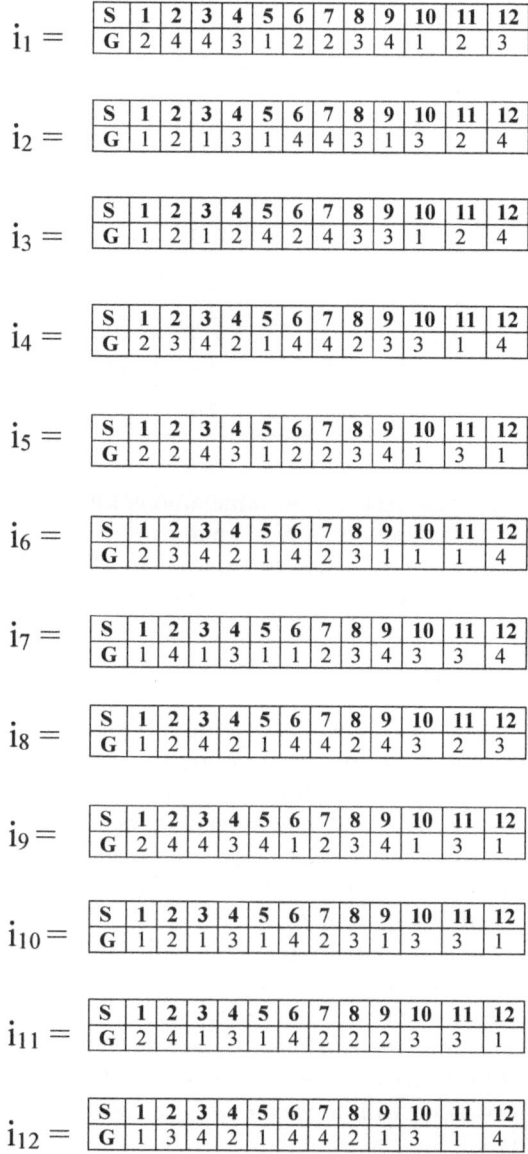

FIGURE 3.1: Twelve solutions taken randomly as initial population

3.3.5 Parent selection

Due to the sorting and partitioning procedure, the first solution in each subcomplex is the best, and last is the worst solution in that complex. The best solution is considered as Xb, and the worst solution is considered as Xw.

TABLE 3.4: Partitioning of solutions into p complexes

A_1	i_1	i_4	i_7	i_{10}
A_2	i_2	i_5	i_8	i_{11}
A_3	i_3	i_6	i_9	i_{12}

TABLE 3.5: Partitioning of p complexes into v sub-complexes

	V_1	V_2
A_1	i_1, i_4	i_7, i_{10}
A_2	i_2, i_5	i_8, i_{11}
A_3	i_3, i_6	i_9, i_{12}

Also, the solution with the best fitness value among all solutions is considered as a global solution and denoted as Xg. This process is used to improve the solution with the worst fitness value in each cycle. So the best solution Xb and worst solution Xw are used in next offspring generation phase.

3.3.6 Offspring generation

During the proposed offspring generation, only one offspring is generated by copying the information from the best solution after the crossover point to the worst solution. The best solution remains at its own place, and the worst solution is replaced by the new offspring. In the next iteration, this newly generated offspring is treated as the worst solution. This process helps to retain the best solution. The overall process of new offspring generation for each complex is shown in Figure 3.2.

From Table 3.5, in sub-complex (A_2, V_2), solution i_8 is considered as Xb and i_{11} is considered as Xw. Here, we have used single-point crossover because it gives better results than double-point crossover and uniform crossover [62]. In our implementation, only one offspring is generated by copying the information of Xb to Xw. The crossover point is selected as 3. Hence, the information after the third sensor node will be exchanged, as shown in Figure 3.3.

We have evaluated the individuals from all the complexes and subcomplexes. This partitioning of an array D into several complexes reduces the time to search the best individual (solution).

Theorem 1 *Offspring produced according to section 3.3.6 is valid.*

Proof 1 *A valid solution is one in which; the sensor node is assigned to a gateway which is in range of respective sensor node. During the initial population phase, sensor nodes are assigned to the gateway within communication*

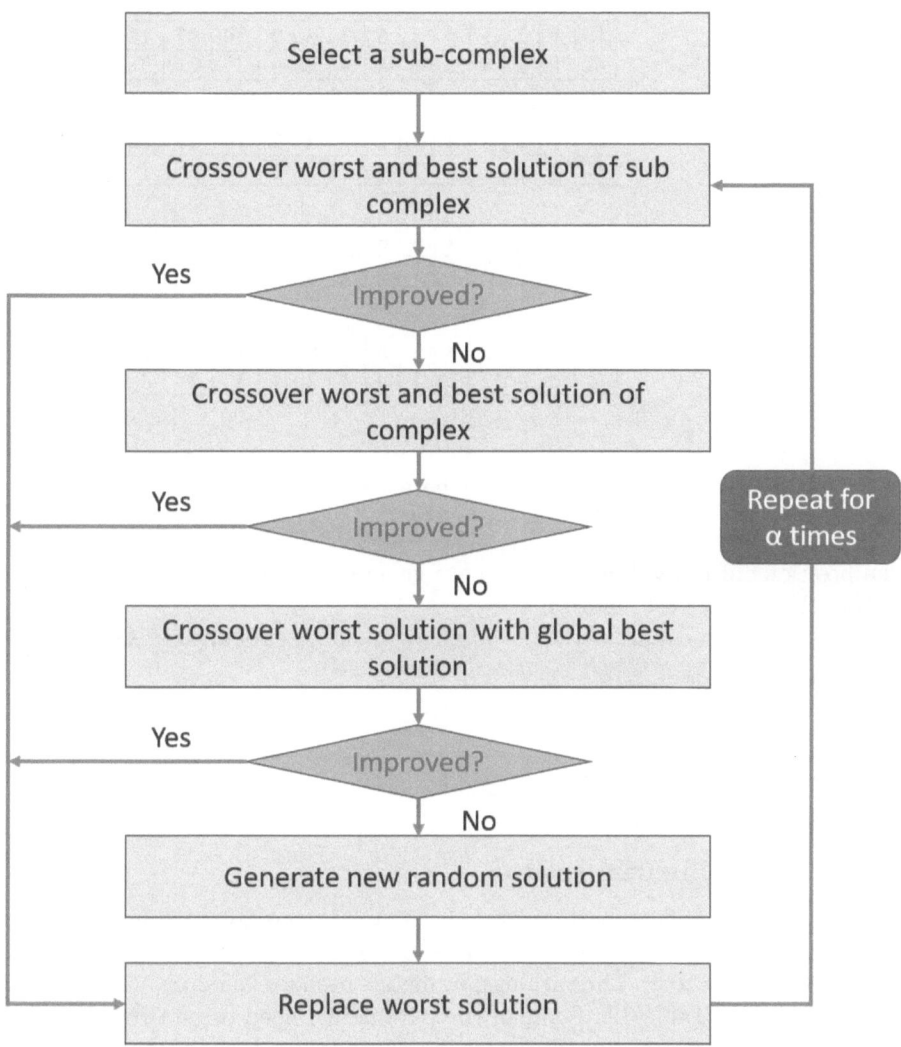

FIGURE 3.2: Steps in evaluation of a single complex

range only. So, at the time of offspring generation, the gateway associated with the sensor node is one of the gateways in the range of the respective sensor node. Hence, for every sensor node, its corresponding gateway is valid.

3.3.7 Sorting and shuffling

All the solutions are sorted and stored again in an array D with updated fitness value along with their position information. This causes shuffling in

i_8 (Xb) =

S	1	2	3	4	5	6	7	8	9	10	11	12
G	1	2	4	2	1	4	4	2	4	3	2	3

i_{11} (Xw) =

S	1	2	3	4	5	6	7	8	9	10	11	12
G	2	4	1	3	1	4	2	2	2	3	3	1

a)

Xb =

S	1	2	3	4	5	6	7	8	9	10	11	12
G	1	2	4	2	1	4	4	2	4	3	2	3

Offspring =

S	1	2	3	4	5	6	7	8	9	10	11	12
G	2	4	1	2	1	4	4	2	4	3	2	3

b)

FIGURE 3.3: Offspring generation: a) Before; b) After

the positions of individuals. In the next generation, the procedure repeats to obtain the global best solution.

This whole procedure of the proposed algorithm is explained in Algorithm 1.

3.4 Results and discussion

3.4.1 Experimental setup

We performed the experiments under a WSN scenario with the number of sensor nodes and gateways positioned in 50×50 m^2 area. The sensing range of the sensor node is 10 m. The various parameters used for the simulation purpose are described in Table 3.6. Some of the parameters need to be tuned finely to obtain the optimal solution. These parameters are also listed in Table 3.7.

TABLE 3.6: Simulation parameters for SCE

Parameter	Value
Area	$50 \times 50 \ m^2$
Base Station Location	$(48, 25)$
Communication Range	$10m$
E_{elec}	$50 \ nJ/bit$
ϵ_{fs}	$10 \ pJ/bit/m^2$
ϵ_{mp}	$0.001 \ pJ/bit/ \ m^4$
Number of sensors	50, 100, 150
Number of gateways	5, 11, 19

TABLE 3.7: Tuning parameter values for SCE initialization

Parameter	Value	Explanation
PopSize	30	Initial population size
α	10	Number of generations for offspring
β	400	Number of generations for solution
γ	0.25	Threshold constant
Number of complexes	2 to 10	Number of partitions in D
Number of sub-complexes	2 to 5	Number of sub-partitions in each complex

To compare the performance of the proposed algorithm, we have implemented some of the existing algorithms such as Simple GA Load Balancing (SGALB) [63], Node Local Density-based Load Balancing NLDLB [14], Novel GA Load Balancing (NGALB) [19] and Score Based Load Balancing (SBLB) [15] in MATLAB. The performance of these algorithms is evaluated under different factors such as fitness values of the solution, energy consumption until the first gateway dies, the round of death of the first gateway, the round number of the death of first sensor node and half of the sensor nodes.

3.4.2   Number of sensor nodes vs energy consumed

The sensor nodes consume energy for receiving and transmitting the data. Here, the experiments have been conducted for three scenarios with the number of sensor nodes 50, 100 and 150. The experiments are also conducted for the equal and unequal load of the sensor nodes. The proposed approach is compared with NGALB, SGALB, NLDLB and SGLB. The energy consumption until the first gateway die is calculated. The experimental results of energy consumption for equal and unequal load are shown in Figure 3.4 (a) and (b), respectively. It is observed from the results that proposed SCE-based approach outperforms compared algorithms. This is due to the algorithm that selects the solution according to the best fitness value. The fitness function contemplates the residual energy of gateways to prolong the lifetime. Therefore, the selected solution is an energy-efficient solution.

3.4.3   Number of heavy loaded sensor nodes vs fitness

For this evaluation parameter, the experiments are conducted for 50, 100 and 150 number of sensor nodes by varying the amount of heavy loaded sensor nodes from 1 to 5. The proposed approach is compared with the NLDLB, SBLB, SGALB, NGALB algorithms. Figure 3.5 (a), (b) and (c) shows the results for 50, 100 and 150 sensor nodes with the variable number of heavy loaded sensor nodes, respectively. From Figure 3.5, it is observed that the proposed SCE approach performs better than other algorithms in all three scenarios. This is because the solution chosen in the proposed method considers the heavy and underloaded gateways. So the algorithm says the best optimal solution is the most balanced solution.

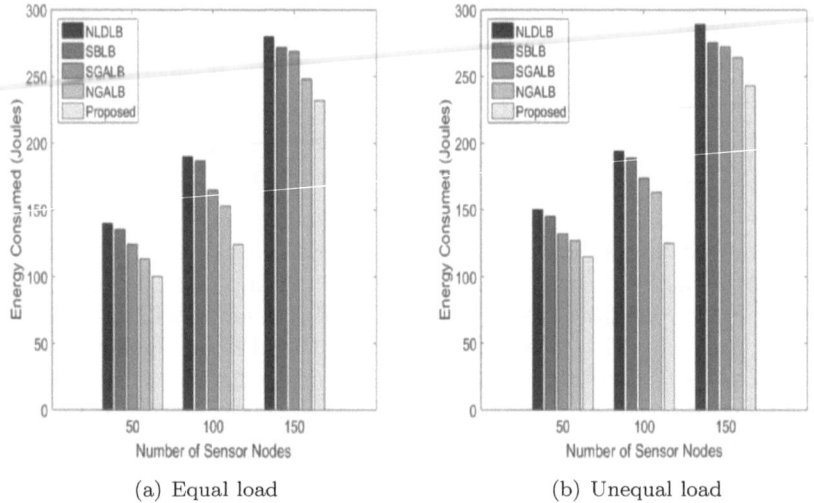

(a) Equal load   (b) Unequal load

FIGURE 3.4: Comparison of the proposed algorithm with NLDLB, SBLB, SGALB and NGALB algorithms in terms of number of sensors and energy consumption for: (a) Equal load; (b) Unequal load

3.4.4   Number of generations vs fitness

In this evaluation factor, the fitness of solution over the number of generations of 200 and 400 is calculated. The SGALB and NGALB algorithms are compared for this parameter. We did not consider NLDLB and SBLB algorithms as they are heuristics approaches. The experiments are performed for 50, 100 and 150 sensor nodes for an equal and unequal load of sensor nodes. The fitness of the solution generated by each algorithm is obtained. The results of these algorithms are shown in Figure 3.6. Figure 3.6 (a) and Figure 3.6 (b) shows the results using 50 sensor nodes for equal and unequal load, respectively. Figure 3.6 (c) and Figure 3.6 (d) shows the results using 100 sensor nodes for equal and unequal load, respectively. Figure 3.6 (e) and Figure 3.6 (f) shows the results using 150 sensor nodes for equal and unequal load, respectively. It is observed to that at each generation proposed, SCE has improved fitness than other compared algorithms. This is because our SCE-based approach carries the best solution forward in the next generation. Hence, there is no loss of optimal solution at each generation.

3.4.5   First node die

This metric defines the number of rounds when the death of the first sensor node occurs. We have also conducted this metric for an equal and unequal load of 150 sensor nodes in Figure 3.7 (a) and (b), respectively. It is observed that

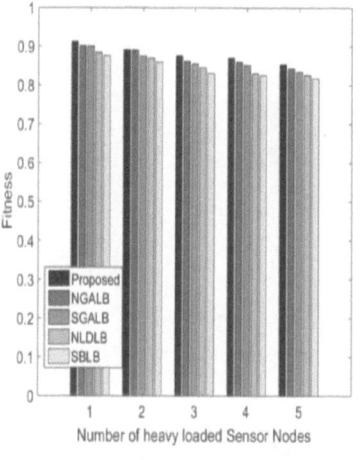

(a) Network with 50 sensors

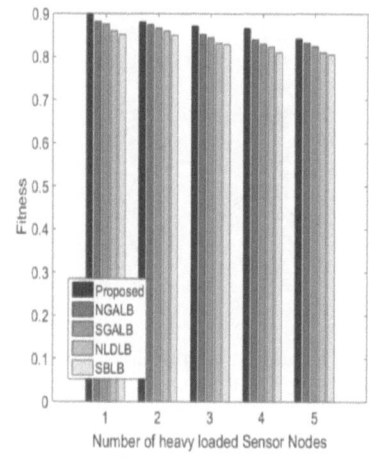

(b) Network with 100 sensors

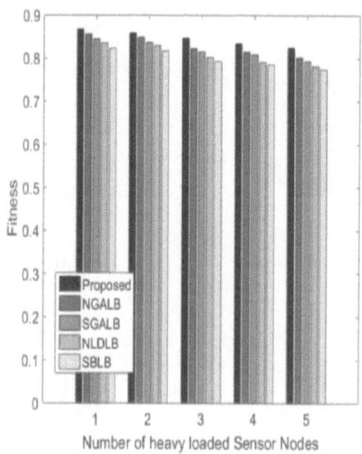

(c) Network with 150 sensors

FIGURE 3.5: Comparison of the proposed algorithm with NLDLB, SBLB, SGALB and NGALB algorithms in terms of number of heavy loaded sensor nodes and fitness value for: (a) Network with 50 sensor nodes, (b) Network with 100 sensor nodes and (c) Network with 150 sensor nodes

the proposed SCE-based approach outperforms NGALB, SGALB, NLDLB and SBLB. The energy consumption rate at the proposed approach is less than other compared algorithms. Therefore, the first node in the proposed SCE-based approach dies after a long time, whereas in other algorithms it dies quickly.

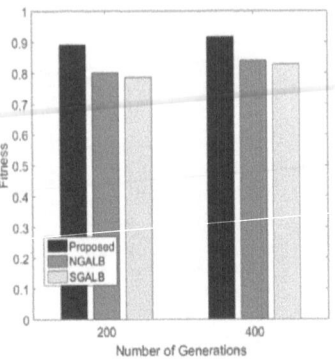

(a) Equal load with 50 sensors

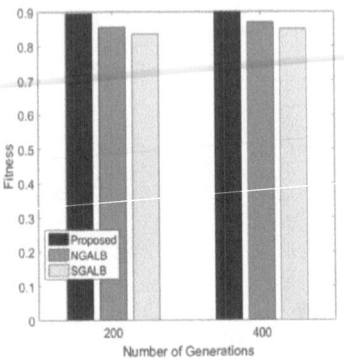

(b) Unequal load with 50 sensors

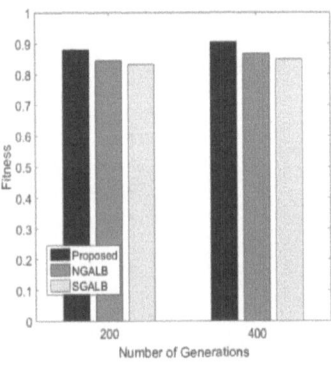

(c) Equal load with 100 sensors

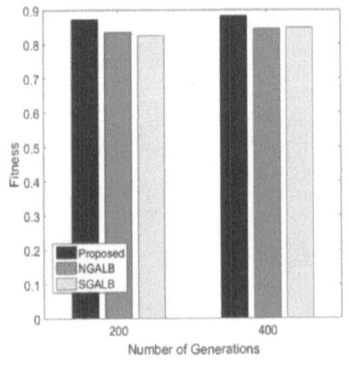

(d) Unequal load with 100 sensors

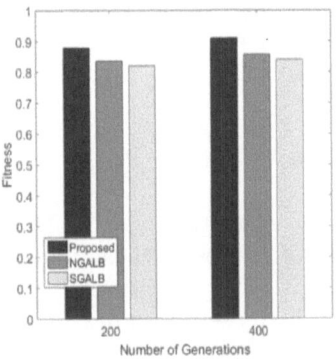

(e) Equal load with 150 sensors

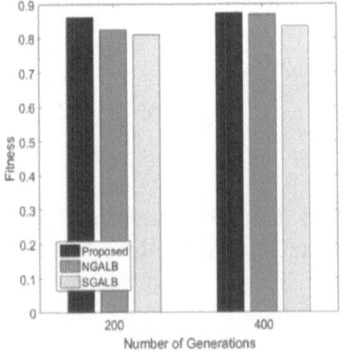

(f) Unequal load with 150 sensors

FIGURE 3.6: Comparison of the proposed method with SGALB and NGALB in terms of number of generations and fitness value for (a) Equal load with 50 sensor nodes, (b) Unequal load with 50 sensor nodes, (c) Equal load with 100 sensor nodes, (d) Unequal load with 100 sensor nodes, (e) Equal load with 150 sensor nodes and (f) Unequal load with 150 sensor nodes

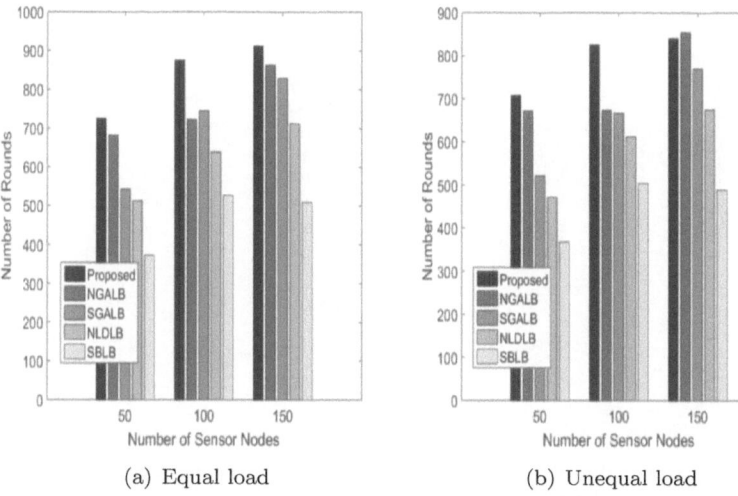

(a) Equal load   (b) Unequal load

FIGURE 3.7: Comparison of the proposed algorithm with NGALB, SGALB, NLDLB and SBLB algorithms in terms of first node die for: (a) Equal load; (b) Unequal load

3.4.6   Half of the nodes alive

The metric half of the nodes alive estimates the rate of the energy consumption of the network. The difference between the death of the first sensor node and the death of the half of the sensor nodes determines the stability of the network. Figure 3.8 (a) and (b) shows the number of rounds at which half of the sensor node is live for 150 sensor nodes and an equal and unequal load of sensor nodes, respectively. It is observed that the proposed SCE-based algorithm outperforms NGALB, SGALB, NLDLB and SBLB algorithms. The sensor nodes die early in the case of compared algorithms than the proposed approach.

3.4.7   First gateway die

This metric defines the network lifetime of the algorithm. It is the round at which the death of the first gateway occurs in the network. Figure 3.9 (a) and (b) shows the comparison of the first gateway die for 150 sensor nodes and the equal and unequal load of sensor nodes, respectively. It is observed that in both cases, the proposed SCE-based algorithm shows a large number of rounds than other compared algorithms.

3.4.8   Number of dead sensor nodes

We have noted the rounds at which 10, 20, 30, 40 and 50 number of sensor nodes die in the network. Figure 3.10 (a) and (b) shows the comparison of

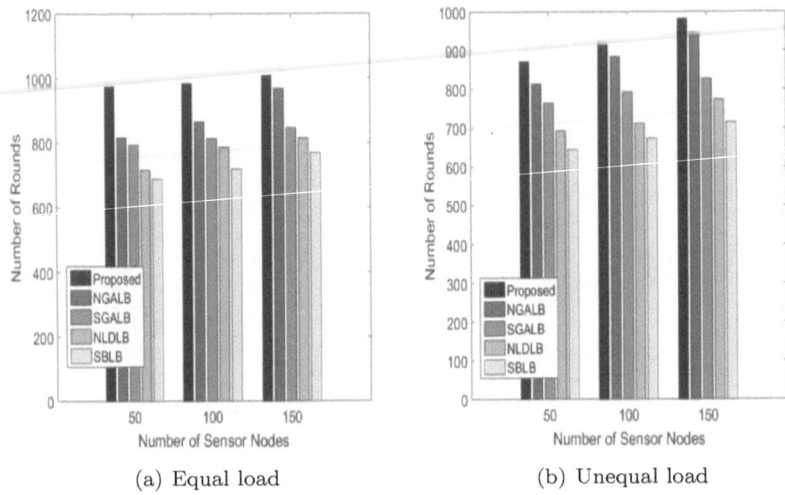

(a) Equal load   (b) Unequal load

FIGURE 3.8: Comparison of the proposed algorithm with NGALB, SGALB, NLDLB and SBLB algorithms in terms of half of the nodes dying for: (a) Equal load; (b) Unequal load

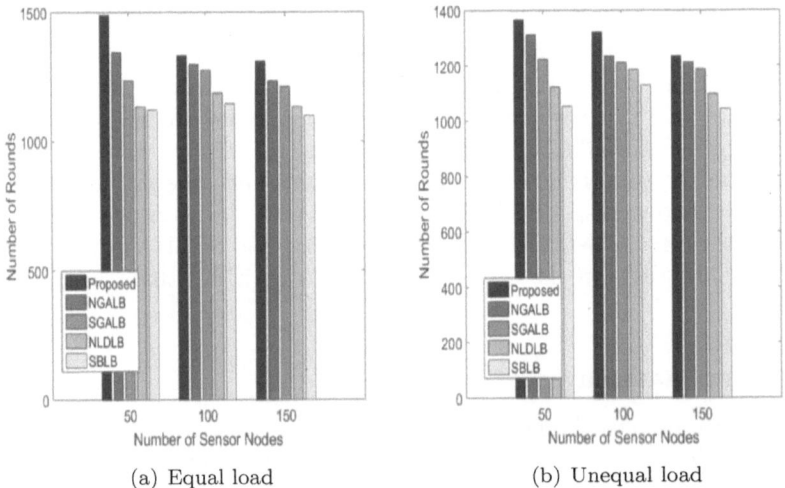

(a) Equal load   (b) Unequal load

FIGURE 3.9: Comparison of the proposed algorithm with NGALB, SGALB, NLDLB and SBLB algorithms in terms of first gateway dying for: (a) Equal load; (b) Unequal load

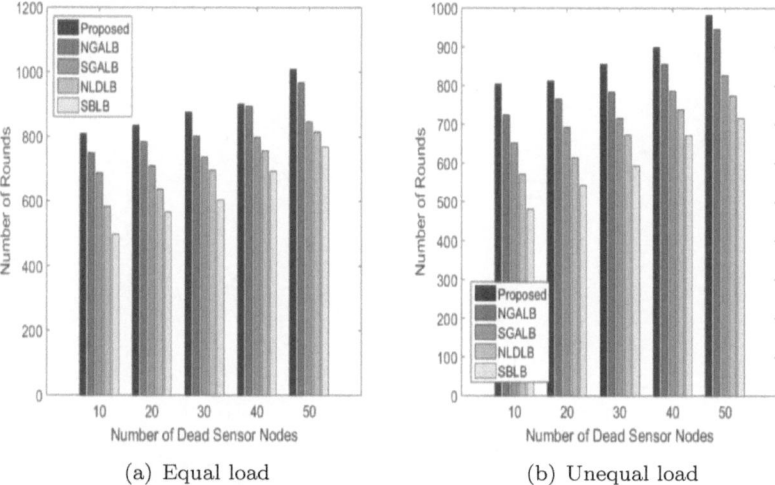

(a) Equal load   (b) Unequal load

FIGURE 3.10: Comparison of the proposed algorithm with NGALB, SGALB, NLDLB and SBLB algorithms in terms of dead sensor nodes for: (a) Equal load; (b) Unequal load

the dead sensor nodes for 150 sensor nodes and an equal and unequal load of sensor nodes, respectively. It is clear from the figure that at each case, proposed SCE outperforms NGALB, SGALB, NLDLB and SBLB algorithms. This is because the proposed SCE-based algorithm always selects a stable solution.

3.5 Conclusion

The SCE approach is applied for load balancing of gateways in WSN. A novel fitness function is designed to measure the quality of the solution. The fitness function is designed according to the heavy and underloaded gateways. A new approach is followed for the generation of offspring, that the better offspring replaces only the worst solution and takes over the best solutions in the next generation. The performance of the proposed approach is compared with state-of-the-art load balancing algorithms. It is observed that the proposed algorithm outperforms these algorithms.

Chapter 4

Novel Fitness Function for SCE Algorithm Based Energy Efficiency in WSN

4.1 Introduction

In this chapter, we focus on energy efficiency with load balancing using Shuffled Complex Evolution (SCE) with some improvements called ISCE. As an improvement, two genetic operators, namely crossover and mutation, have included generating healthy offsprings. A novel fitness function is designed to measure the quality of the solutions. The experiments are conducted under various evaluation factors, where ISCE is found to be an energy-efficient algorithm.

4.2 Proposed algorithm

The various improvements applied to SCE [58] are discussed below.

4.2.1 Research contribution

1. In the phase of the initial population, the assignment of the sensor node and gateway is restricted to their communication range, instead of random assignment. So that, in the first phase itself, valid solutions are generated.

2. Instead of producing two offsprings, only a single offspring is generated to preserve the best solution from the previous generation.

3. An inventive fitness function is designed to calculate the excellence of the solution generated by the algorithm.

4. We have also added a new phase to the SCE called relocation phase. In the relocation phase, the sensor node with the highest distance from any of the gateways is assigned to its nearest gateway to save the energy of the sensor node.

The proposed ISCE algorithm is discussed in details below:

4.2.2   Initial population generation

In the proposed ISCE algorithm, the solution is constructed with an assignment of gateways and the sensor nodes within its communication range. Let us study the following Example 1 to understand the initial population generation phase as follows.

Example 1: Consider a WSN with the set of 6 sensor nodes $S = \{s_1, s_2, \ldots, s_6\}$ and set of 3 gateways $G = \{g_1, g_2, g_3\}$. The size of the solution is equal to the number of sensor nodes in the network. The gateways in the communication range of every sensor node are shown in Table 4.1. Each sensor is allocated to the gateway from their respective gateway list only. Table 4.2 shows the valid solution for the WSN mentioned in Example 1. Sensor node s_1 is assigned to gateway g_1, sensor node s_2 is assigned to g_3, and so on.

4.2.3   Proposed fitness function

A fitness function is used to test the excellence of the solution. Here, we have designed a novel fitness function to balance the gateway load with respect to the distance of the gateway from the BS. The Equation 4.1 illustrates proposed energy-efficient fitness function with the parameters as distance ratio, the lifetime of gateway [19] and projected gateway load.

$$Fitness = \sum_{i=1}^{m} Distance\ Ratio\ (G_i) * \sum_{i=1}^{m} Lifetime\ (G_i)$$
$$* \ Projected\ gateway\ load \tag{4.1}$$

TABLE 4.1: List of gateways within communication range (R) of sensor nodes

Sensor Nodes	Gateways in Range
s_1	g_1, g_2
s_2	g_2, g_3
s_3	g_1
s_4	g_1, g_3
s_5	g_2
s_6	g_1, g_2, g_3

TABLE 4.2: Example of valid solution from initial population

S	1	2	3	4	5	6
G	1	3	1	3	2	2

Distance Ratio (G_i) in Equation 4.2 is termed as overall distance traversal of all the gateways towards BS.

$$Distance\ Ratio\ (G_i) = \frac{Dist\ (G_i, BS)}{max(Dist(G_i, BS))\forall i = 1, 2, \ldots, m} \tag{4.2}$$

where m is total number of gateways in the network.

$$Lifetime(G_i) = \frac{E_{remaining}}{E_{gateway}} \tag{4.3}$$

The lifetime of the gateway is defined as the ratio of remaining energy of gateway and initial energy of gateways [19] as expressed in Equation 4.3. The average gateway load on each gateway is the projected gateway load and formulated in Equation 4.4.

$$Projected\ gateway\ load = \frac{\sum_{i=1}^{m} Load\ Ratio\ (G_i)}{Number\ of\ Gateways} \tag{4.4}$$

Again load ratio is calculated similar to the distance ratio and defined in Equation 4.5

$$Load\ Ratio\ (G_i) = \frac{Load\ (G_i)}{max(Load(G_i))\forall i = 1, 2, \ldots, m} \tag{4.5}$$

$Load(G_i)$ is the number of data bits received by the gateway G_i from sensor nodes as well as gateways from routing path.

4.2.4   Sorting and partitioning of complexes

From Example 1, the initial eight solutions are evaluated according to the fitness function and sorted in a descending manner according to the fitness values as presented in Figure 4.1.

Moreover, according to step 4 and step 5 from SCE algorithm, the partitioning of solutions is conducted. The set of solutions is divided into the two complexes, and again each complex is divided into two sub-complexes. Figure 4.2 shows this partitioning among complexes and sub-complexes.

A sub-complex is represented as p_i, q_j where i and j are the complex number and sub-complex number, respectively.

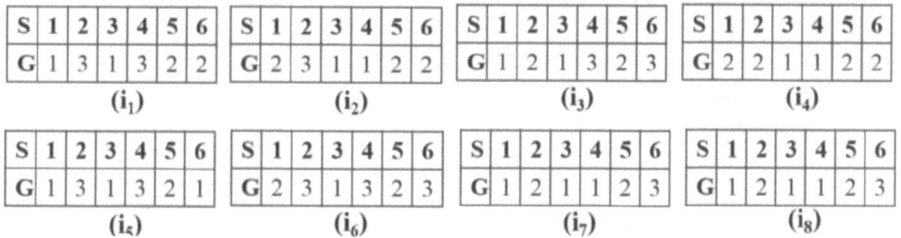

FIGURE 4.1: Initial population with eight solutions

	q_1	q_2	q_1	q_2
p_1	i_1	i_3	i_5	i_7
p_2	i_2	i_4	i_6	i_8

FIGURE 4.2: Partitioning of eight solutions in complexes and sub-complexes

4.2.5 Selection of parent and offspring generation

The selection of parents is conducted as the first and last solution from the sub-complexes or the complexes. It depends on the fitness value obtained from new offspring. By continuing Example 1 and considering Figure 4.2, for understanding purposes, here we select sub-complex (p_2, q_2). i_6 and i_8 are best and worst solutions from sub-complex (p_2, q_2), respectively. We apply a one-point crossover operation [62] by exchanging the information between solution i_6 and solution i_8 to generate new offspring.

Figure 4.3 shows the newly generated offspring and the worst solution i_8 is replaced by newly created offspring.

Lemma: *The solution obtained from crossover operation is valid.*
Proof: The assignment of sensor nodes is carried out in the initial population phase by choosing a gateway in its communication range, therefore, in the phase of crossover; also, the gateway exchanged is from its communication range only. Therefore, the solution generated at the crossover operation is valid.

4.2.6 Relocation phase

Let us assume that, in newly generated offspring from Figure 4.3, sensor node s_2 is farther from gateway g_2; then the assignment of the sensor node s_2 is replaced by its nearer gateway. And let us assume gateway g_3 is nearer to sensor node s_2; then assign s_2 to g_3 as shown in Figure 4.4. This is called relocation.

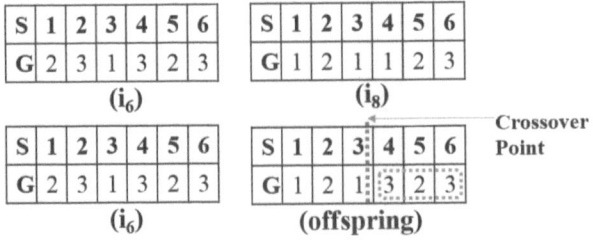

FIGURE 4.3: Offspring generation

S	1	2	3	4	5	6
G	1	3	1	3	2	3

(off-spring after relocation phase)

FIGURE 4.4: Offspring generated after relocation phase

4.2.7 Sorting and shuffling

After generation of offsprings, they are evaluated by the fitness function. All the solutions are then sorted in descending order, and this causes shuffling in the solutions.

4.3 Results and discussion

4.3.1 Experimental setup

The experiments are performed under various scenarios with 50, 100 and 150 number of sensor nodes and the diverse number of gateways deployed in 50×50 m^2 area. The communication range of the sensor nodes is considered as 10 m. The various parameters used for the simulation purpose are described in Table 4.3. Some of the parameters need to be tuned finely to obtain the optimal solution. These parameters are also listed in Table 4.4.

TABLE 4.3: Simulation Parameters for ISCE

Parameter	Value
Area	50×50 m^2
Base Station Location	$(48, 25)$
Communication Range	$10m$
E_{elec}	$50 \ nJ/bit$
ϵ_{fs}	$10 \ pJ/bit/m^2$
ϵ_{mp}	$0.001 \ pJ/bit/ \ m^4$
Number of sensor nodes	50, 100, 150
Number of gateways	5, 11, 19

TABLE 4.4: Tuning parameter values for ISCE initialization

Parameter	Value	Explanation
PopSize	30	Initial population size
α	10	Number of generations for offspring
β	400	Number of generations for solution
γ	0.25	Threshold constant
Number of complexes	2 to 10	Number of partitions in D
Number of sub-complexes	2 to 5	Number of sub-partitions in each complex

To compare the performance of the proposed algorithm, we have implemented existing Node Local Density-based Load Balancing NLD [14], Novel GA Load Balancing (NGALB) [19] and Score Based Load Balancing (SB) [15] algorithms in MATLAB.

The performance of these algorithms are evaluated under different factors such as fitness values of the solution, energy consumption until the first gateway dies, the round of death of the first gateway, the round number of the death of first sensor node and half of the sensor nodes.

4.3.2 Number of sensor nodes vs energy consumed

The energy consumed by sensor nodes and gateways is calculated. The sensor nodes require energy to receive and transmit the data from one sensor node or gateway to another. The experiments are conducted for all three scenarios and with an equal and unequal load of sensor nodes. The proposed approach is compared with NGALB, SGALB, NLD and SB. The energy consumption until the first gateway die is calculated is noted. The experimental results of energy consumption for equal and unequal load are shown in Figure 4.5 (a) and (b), respectively. It is seen that the proposed ISCE approach outperforms compared algorithms. It is due to the selection of energy-efficient solutions according to the fitness function. The fitness function contemplates the residual energy of gateways in order to prolong the lifetime.

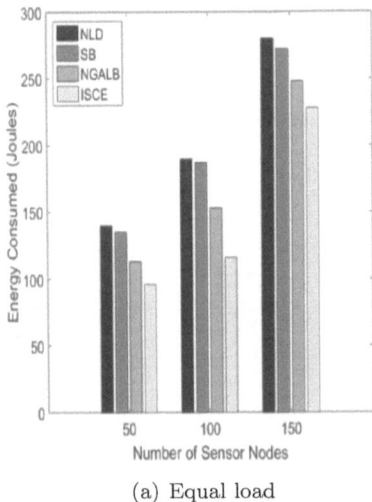

(a) Equal load

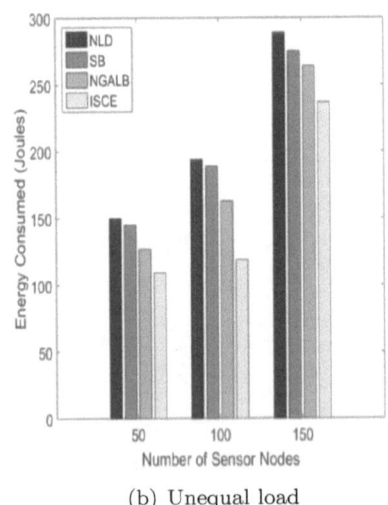

(b) Unequal load

FIGURE 4.5: Comparison of the proposed ISCE with NLD, SB and NGALB algorithms in terms of number of sensors and energy consumption for: (a) Equal load; (b) Unequal load

4.3.3 Number of heavy loaded sensor nodes vs fitness

In this evaluation factor again, the experiments are performed for the variable load of 50, 100 and 150 number of sensor nodes. The load is varied from one packet to five packets at a time. The proposed approach is compared with the NLD, SB, SGALB, NGALB algorithms. Figure 4.6 (a), (b) and (c)

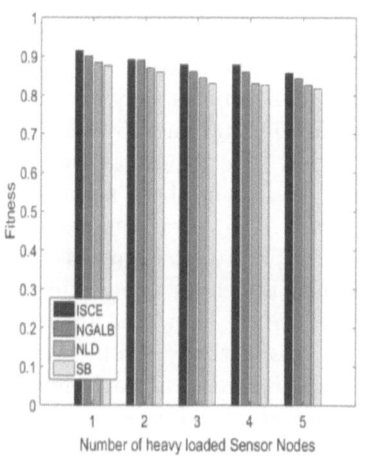

(a) Network with 50 sensors

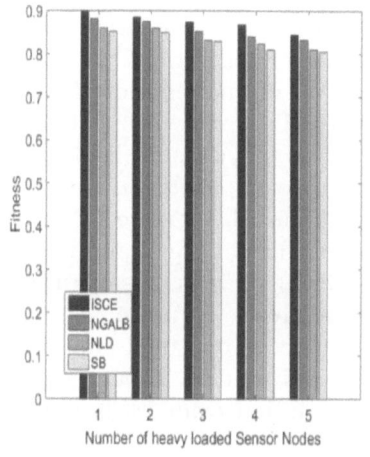

(b) Network with 100 sensors

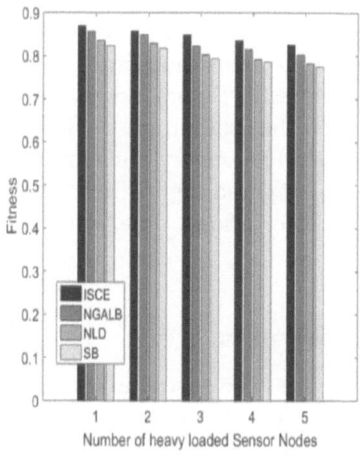

(c) Network with 150 sensors

FIGURE 4.6: Comparison of the proposed ISCE with NLD, SB and NGALB algorithms in terms of number of heavy loaded sensor nodes and fitness value for: (a) Network with 50 sensor nodes, (b) Network with 100 sensor nodes, and (c) Network with 150 sensor nodes

shows the results for 50, 100 and 150 sensor nodes with the variable number of heavily loaded sensor nodes, respectively. From Figure 4.6, it is observed that the proposed ISCE approach performs better than other algorithms in all three scenarios. This is because, at each iteration, the load-balanced solution is chosen as the best solution.

4.3.4   Number of generations vs fitness

For this factor of evaluation, we have considered the solutions over 200 and 400 number of generations. The proposed ISCE algorithm is compared with NGALB. We did not consider NLD and SB algorithms as they are heuristics approaches. The experiments are performed for 50, 100 and 150 sensor nodes for an equal and unequal load of sensor nodes. The fitness of the solution obtained by each algorithm is obtained. The results of these algorithms are shown in Figure 4.7. Figure 4.7 (a) and Figure 4.7 (b) show the results using 50 sensor nodes for equal and unequal load, respectively. Figure 4.7 (c) and Figure 4.7 (d) show the results using 100 sensor nodes for equal and unequal load, respectively. Figure 4.7 (e) and Figure 4.7 (f) show the results using 150 sensor nodes for equal and unequal load, respectively. From the results, it is observed that ISCE outperforms NGALB, because, at each generation, the best solution from ISCE is preserved and carried forward to the next generation.

4.3.5   First node die

The first node dies metric defines the time (round) of the death of the first sensor node. The experiments for this metric are conducted for an equal and unequal load of 150 sensor nodes in Figure 4.8 (a) and (b), respectively. From the results obtained, it is seen that the proposed SCE-based approach outperforms NGALB, NLD and SB. The energy consumption rate at the proposed approach is less than other compared algorithms. Due to which, the death of the first sensor node occurs after a long time, whereas in other algorithms it dies quickly.

4.3.6   Half of the nodes alive

This evaluation factor defines the sustainability period of the network and estimates the rate of the energy consumption of the network. The difference between the death of the first sensor node and the death of half of the sensor nodes determines the stability of the network. Figure 4.9 (a) and (b) shows the number of rounds at which half of the sensor node is live for 150 sensor nodes and an equal and unequal load of sensor nodes, respectively. It is observed that the proposed SCE-based algorithm outperforms NGALB, NLD and SB algorithms. The sensor nodes die early in the case of compared algorithms than the proposed approach.

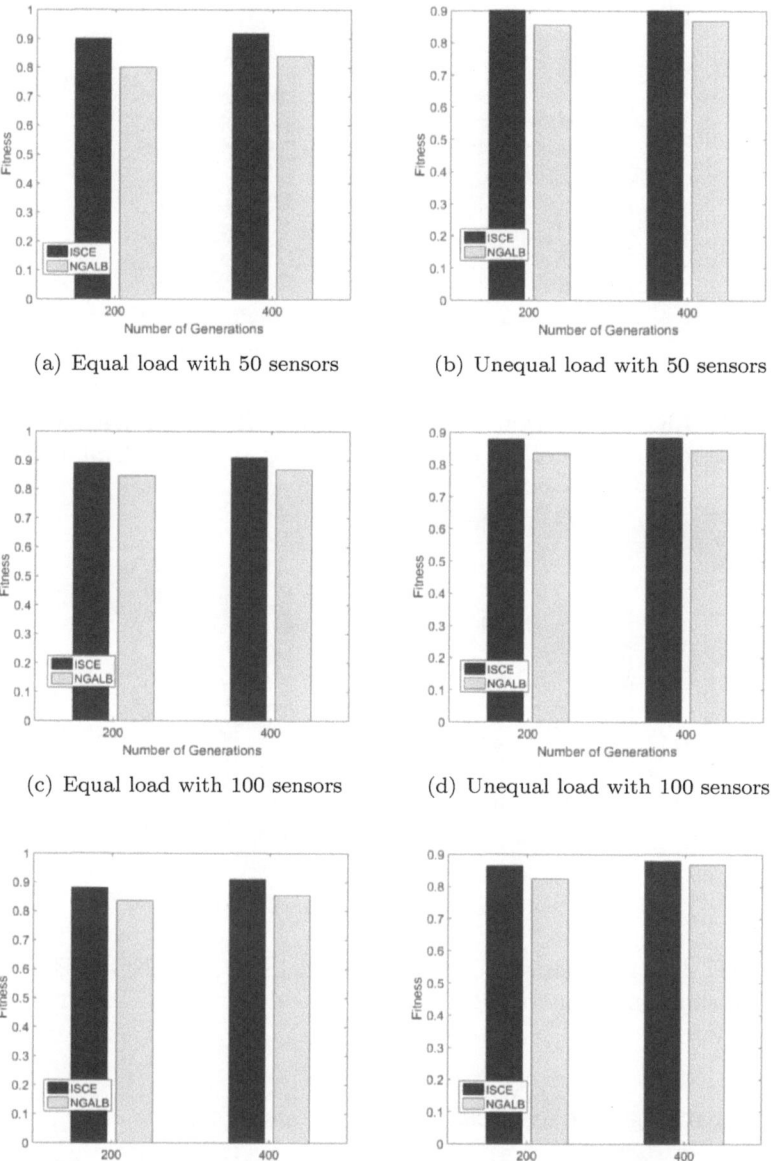

(a) Equal load with 50 sensors

(b) Unequal load with 50 sensors

(c) Equal load with 100 sensors

(d) Unequal load with 100 sensors

(e) Equal load with 150 sensors

(f) Unequal load with 150 sensors

FIGURE 4.7: Comparison of the proposed ISCE with NGALB in terms of number of generations and fitness value for: (a) Equal load with 50 sensor nodes, (b) Unequal load with 50 sensor nodes, (c) Equal load with 100 sensor nodes, (d) Unequal load with 100 sensor nodes, (e) Equal load with 150 sensor nodes, and (f) Unequal load with 150 sensor nodes

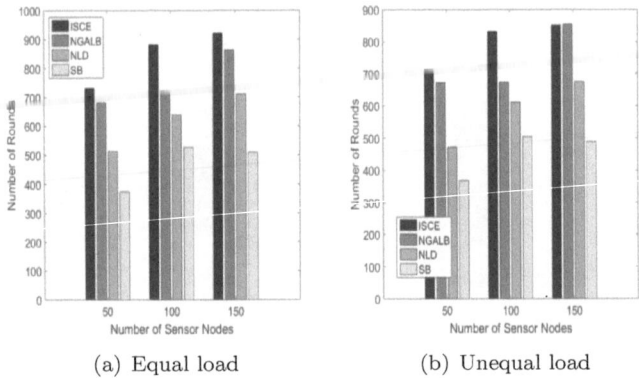

(a) Equal load   (b) Unequal load

FIGURE 4.8: Comparison of the proposed ISCE with NGALB, NLD and SB algorithms in terms of the death of first node: (a) Equal load; (b) Unequal load

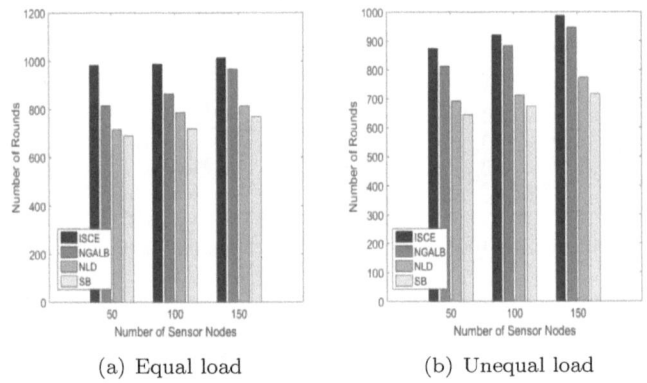

(a) Equal load   (b) Unequal load

FIGURE 4.9: Comparison of the proposed ISCE with NGALB, NLD and SB algorithms in terms of half of the nodes dying for: (a) Equal load; (b) Unequal load

4.3.7   First gateway death

The metric first gateway die indicates the lifetime of the network. Figure 4.10 (a) and (b) shows the comparison of the first gateway die for 150 sensor nodes and the equal and unequal load of sensor nodes, respectively. It is observed that in both cases, the proposed ISCE algorithm shows a large number of rounds than NGALB, NLD and SB algorithms.

4.3.8   Number of dead sensor nodes

We have noted the rounds at which 10, 20, 30, 40 and 50 number of sensor nodes die in the network. Figure 4.11 (a) and (b) shows the comparison of the

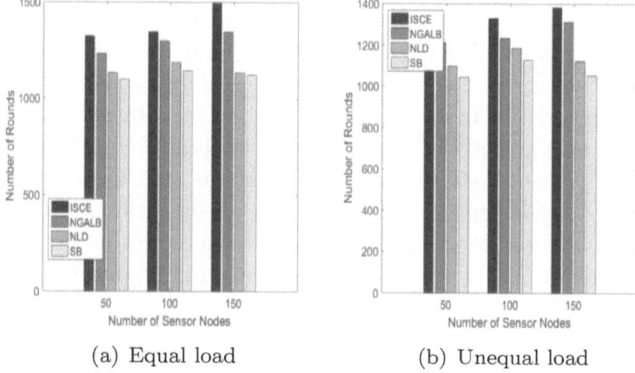

(a) Equal load   (b) Unequal load

FIGURE 4.10: Comparison of the proposed ISCE with NGALB, NLD and SB algorithms in terms of first gateway death for: (a) Equal load; (b) Unequal load

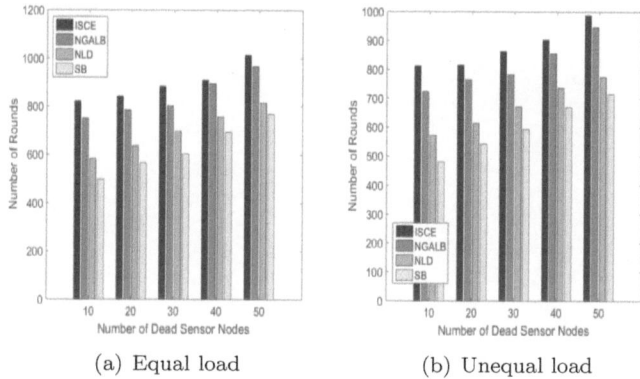

(a) Equal load   (b) Unequal load

FIGURE 4.11: Comparison of the proposed ISCE with NGALB, NLD and SB algorithms in terms of dead sensor nodes for: (a) Equal load; (b) Unequal load

dead sensor nodes for 150 sensor nodes and an equal and unequal load of sensor nodes, respectively. It is clear from the figure that at each case, proposed ISCE outperforms NGALB, NLD and SB algorithms. This is because the proposed ISCE algorithm always selects a stable solution.

4.4 Conclusion

In this chapter, various improvements in the Shuffled Complex Evolution algorithm are applied to balance the load of gateways in WSNs. ISCE varies from SCE in terms of phases in initial population generation and generation of offspring phase. A new phase has been added after offspring generation phase called relocation phase. The relocation phase helps in minimizing the energy consumption of sensor nodes. Along with these modifications, a novel fitness function is designed for measuring the quality of solution in terms of load ratio, distance ratio and a lifetime of gateways. The ISCE algorithm has been compared with existing load balancing algorithms considering the parameters such as heavily loaded sensor nodes, energy consumption, load balancing for both equal and unequal load, and network lifetime parameters. It is observed that our ISCE algorithm gives out-performance in the parameters mentioned above.

Chapter 5

An Efficient Load Balancing of Gateways Using Improved SFLA for WSNs

5.1 Introduction

Shuffled Frog Leaping Algorithm (SFLA) is an evolutionary algorithm used for local and global search optimization. The SFLA method is a memetic meta-heuristic algorithm developed for solving combinatorial optimization problems. The local search is completed using a particle swarm optimization-like method adapted for discrete problems but emphasizing a local search. To ensure global exploration, the virtual frogs are periodically shuffled and reorganized into new memeplexes in a technique similar to that used in the shuffled complex evolution algorithm. Also, to provide the opportunity for random generation of improved information, random virtual frogs are generated and substituted in the population.

In this chapter, we discuss how to use the SFLA Method efficiently and effectively for energy efficiency in the wireless sensor network. The competent local search feature of SFLA algorithm is suitable for solving different optimization problems. Energy efficiency is one of the challenges in WSN. A novel fitness function is discussed to measure the effectiveness of the generated solution. The solution is represented as the assignment of the sensor node and gateways.

5.2 Preliminaries

5.2.1 An overview of shuffled frog leaping algorithm

Shuffled Frog Leaping Algorithm (SFLA) [64] is defined in terms of global exploration and local exploration.

Global Exploration:

Step 1. **Initialization:**
Select two numbers m and f, where m is number of memeplexes and f is number of frogs in every memeplex. So, the total sample size F in swamp is given in Equation 5.1

$$F = m * f \qquad (5.1)$$

Step 2. **Generation of virtual population:**
Sample F virtual frogs $U(1), U(2), \ldots, U(F)$ in feasible space $\Omega \subset \Re^d$, where d is the number of decision variable. The ith frog is denoted as a vector of decision variable values $U(i) = (U_{i1}, U_{i2}, \ldots, U_{di})$. Calculate the fitness value $f(i)$ for every frog $U(i)$.

Step 3. **Ranking frogs:**
Sort all F frogs in decreasing order with their fitness value. Store them in array X, where $X = U(i), f(i), i = 1, \ldots, F$, therefore $i = 1$ represents the frog with the best fitness value. Let P_X be the best frog's position in the entire population (where $P_X = U(1)$).

Step 4. **Partitioning of frogs into memeplexes:**
Partition array X into m memeplexes $M_1, M_2, \ldots, M_m$, each containing k frogs, such that first frog goes to first memeplex, second frog goes to second memeplex, m^{th} frog goes to m^{th} memeplex and $(m+1)^{th}$ frog goes to first memeplex and so on.

Step 5. **Memetic evolution within each memeplex:**
Evolve each memeplex $Y_k, k = 1, \ldots, m$ according to the frog-leaping algorithm as local exploration as explained below.

Step 6. **Shuffling of memeplexes:**
Replace $Y_1, \ldots, Y_m$ into X, after doing a prior number of memetic evolutionary steps in every memeplex in such a way that $X = Y_k, k = 1, \ldots, m$. Sort X in decreasing order with fitness value.

Step 7. **Checking convergence:**
If the convergence criteria are satisfied, stop. Otherwise, return to Step 4. Generally, the loop runs up to a prior number of generations and stops when at least one frog carries the same *"best memetic pattern"* without change.

Local Exploration:
In Step 5 of global exploration, each memeplex evolves N times independently. After the evolution of memeplex, the algorithm returns to the shuffling process. Following are the steps of local search for each memeplex.

Step 1. Start with $m = 0$ where m computes the number of memeplexes and will be compared with total number of m memeplexes. Set $n = 0$ where n counts the number of evolutionary steps and will be compared with the maximum number N of steps to be completed within each memeplex.

Step 2. Increment m by 1.

Step 3. Increment n by 1.

Step 4. **Construction of a sub-memeplex:** The goal of the frog is to travel towards excellent ideas. Frogs adapt ideas from the best frog in their memeplex or from best frog to improve their memes. Generally, frog tends to concentrate around the local optimum frog. Therefore a subset of the memeplex is considered known as sub-memeplex. The submemeplex approach is to provide higher weight to frogs having higher fitness value. The weights are assigned by a triangular probability distribution, as shown in Equation 5.2

$$p_j = 2 * \frac{(n + 1 - i)}{n * (n + 1)} \tag{5.2}$$

Therefore the frog having higher fitness value in submemeplex can be defined as $p_1 = 2/n + 1$ and the frog having lowest fitness value in submemeplex can be defined as $p_n = 2/n(n + 1)$ The submemeplex array N is formed by selecting n distinct frogs randomly from k frogs in each memeplex. Again submemeplex is sorted in descending order of fitness value. The frog with best and worst fitness value in every submemeplex is considered as P_B and P_W, respectively.

Step 5. **Improving the position of worst frog:** The new position of the worst frog from the sub-complex is computed by
for a positive step:
$S = min(int[rand(P_B - P_W)], S_{max})$
for a negative step:
$S = max(int[rand(P_B - P_W)], -S_{max})$

where *rand* is a random number in the range of 0 to 1. S_{max} is the maximum step size allowed to be adopted by frog after being infected. The step size has dimensions equal to the number of decision variables. Equation 5.3 then computes the new position.

$$U_{(q)} = P_W + S \tag{5.3}$$

If $U_{(q)}$ is within the feasible space Ω, compute the new fitness value $f_{(q)}$. Else go to step 6. If the new $f_{(q)}$ is better than the old $f_{(q)}$, i.e., if the evolution produces a benefit, then replace the old $U_{(q)}$ with the new $U_{(q)}$ and go to step 8. Else go to step 6.

Step 6. If step 5 cannot produce better result than old frog, then the step and new position for that frog are computed by
for a positive step:
step size of $S = min(int[rand(P_X - P_W)], S_{max})$
for a negative step:
$S = max(int[rand(P_X - P_W)], -S_{max})$

And the new position is calculated by Equation 5.3. If $U_{(q)}$ is within the feasible space Ω, calculate the new fitness value $f_{(q)}$; else go to step 7. If the new $f_{(q)}$ is better than the old $f_{(q)}$, i.e., if the evolution produces a benefit, then replace the old $U_{(q)}$ with the new $U_{(q)}$ and go to step 8. Else go to step 7.

Step 7. **Censorship (Transfer phase):** If the new position is unrealistic or not better than old position, new frogs are generated randomly at feasible location in order to replace the frog which is unfeasible to progress ahead, and spread of defective meme is stopped. Compute $f_{(r)}$ and set $U_{(q)} = r$ and $f_{(q)} = f_{(r)}$.

Step 8. **Upgrading of memeplex:** After doing all the changes in the worst frog in each sub-memeplex, replace Z in their original locations. Again sort all the frogs with descending order according to their fitness value.

Step 9. If $n < N$, go to step 3.

Step 10. If $m < M$, go to step 2.
Else return to the global search to shuffle memeplexes.

Step 5 and 6 of local exploration are identical to Particle Swarm Optimization (PSO), as it provides a decent direction to frog and frog moves in that direction [64].

5.3 Proposed methodology

The SFLA [64] is improved with various modifications to resolve the problem of load balancing of gateways in WSNs. The improvements regarding this are discussed below:

1. *Reforming the initialization phase:* We have controlled the connection between sensor nodes and gateways with the communication range of gateways and sensor nodes. This restriction produces valid solutions in the first phase itself. The conventional SFLA uses randomization process to initialize the solutions.

2. *Reforming the offspring generation phase:* Generally, two offsprings are generated in SFLA, but only one offspring is produced for our problem statement. This helps to survive the fittest solution in further generations.

3. *Inclusion of new transfer phase:* In the transfer phase, the heavily loaded gateway eliminates the connection of the farthest sensor node from it. The eliminated sensor node then gets connected to its nearest gateway. This helps in overall load balancing of the network and also reduces the energy consumption of gateway and sensor node.

The proposed algorithm is explained in the following subsections:

5.3.1 Individual representation

The individual (frog) represents a string of mapping of gateways to sensor node. The length of every individual is equal to the number of sensor nodes. An example of individual representation is shown in Example 1 and in Figure 5.1. According to the figure it is interpreted as sensor nodes s_4, s_5, s_6 and s_7 being allocated to gateways g_1, g_4, g_3 and g_2, respectively, and s_1, s_3, s_4 being allocated to gateway g_1.

Example 1: Consider a WSN of 4 gateways and 12 sensor nodes, i.e., $G = g_1, g_2, g_3, g_4$ and $S = s_1, s_2, \ldots, s_{12}$. The length of individual is equal to the number of sensor nodes, i.e., 12. Figure 5.1 shows individual representation for this WSN. Here the gateway g_2 is allocated to sensor node s_7. Likewise, s_4, s_5 and s_6 are allocated to g_1, g_4 and g_3, respectively.

5.3.2 Initialization phase

The individuals are said to be valid if and only if a sensor node is connected to a gateway in its communication range. The valid individual is considered as the complete clustering solution. The set of valid individuals is called an initial population. This phase is explained with an example given below:

Example 2: Consider a WSN of $G = \{g_1, g_2, g_3, g_4\}$ and $S = \{s_1, s_2, \ldots, s_6\}$. Table 5.1 shows the gateways in the range of each sensor node. According to Table 5.1, sensor node s_1 is assigned to either gateway g_1 or g_2. A valid solution generated using Table 5.1 is shown in Figure 5.2.

S	1	2	3	4	5	6	7	8	9	10	11	12
G	1	2	1	1	4	3	2	3	4	3	2	4

FIGURE 5.1: Individual representation

TABLE 5.1: Sensor nodes and their corresponding gateways within communication range

Sensor nodes	Gateways in the range
s_1	g_1, g_2
s_2	g_2, g_3, g_4
s_3	g_1, g_4
s_4	g_2, g_3
s_5	g_1, g_4
s_6	g_1, g_2, g_3

S	1	2	3	4	5	6
G	1	2	1	2	4	3

FIGURE 5.2: One of the valid solutions in initial population

5.3.3   Proposed fitness function

To measure the quality of each individual, efficient fitness function is designed. The fitness function is designed such that maximum load of a gateway is reduced and the load of the gateway is balanced. The Equation 5.4 defines the efficient fitness function for load balancing of gateways.

$$Fitness = (1 - \frac{Mean\ load}{Max\ load}) + \frac{Number\ of\ heavy\ and\ underloaded\ gateways}{Total\ number\ of\ gateways} \quad (5.4)$$

The number of heavy and underloaded gateways is calculated by considering a maximum and minimum threshold of a load of gateways explained in Equation 5.5 and Equation 5.6. The gateways with a load exceeding *maximum threshold* are considered as heavily loaded gateways whereas the gateways with a load less than *minimum threshold* are considered as underloaded gateways.

$$Maximum\ threshold = Mean\ load + \gamma \quad (5.5)$$

$$Minimum\ threshold = Mean\ load - \gamma \quad (5.6)$$

The mean is calculated by Equation 5.7 whereas γ is calculated by Equation 5.8. For calculating γ we have considered *Range of gateway load* and *total number of gateways*. It is defined in Equation 5.9

$$Mean\ load = \frac{\sum_{i=1}^{m} Load\ (G_i)}{Number\ of\ gateways} \quad (5.7)$$

$$\gamma = \frac{Range\ of\ gateway\ load}{Total\ number\ of\ gateways} \tag{5.8}$$

$$Range\ of\ gateway\ load = (maximum\ load\ on\ gateway) - \\ (minimum\ load\ on\ gateway) \tag{5.9}$$

The fitness function is designed for load balancing as well as energy efficiency. Therefore, while considering the load of gateways, we are considering the remaining energy of gateways too as shown in Equation 5.10.

$$Load\ (G_i) = L * \frac{Remaining\ energy\ of\ (G_i)}{Initial\ energy\ of\ (G_i)} \tag{5.10}$$

where i is the gateway number and L is the number of packets sent by the gateway.

Remaining energy of each gateway is calculated by subtracting its consumed energy from initial energy.

For a better solution fitness function needs to be minimized, i.e., minimum fitness value indicates best solution. The fitness calculation is explained using Example 3.

Example 3: Consider g_1, g_2, g_3 and g_4 have loads 15, 9, 18 and 22, respectively.

From Equation 5.7, the mean of loads is $= \frac{(15+9+18+22)}{4} = 16$.

According to given loads the minimum and maximum loads are 9 and 22, respectively.

Therefore, range of gateway load using Equation 5.9 is $= 22 - 9 = 13$.

From Equation 5.8 $\gamma = 13/4 = 3.25$.

According to Equation 5.5 and Equation 5.6 maximum and minimum thresholds are:

Maximum threshold $= 16 + 3.25 = 19.25$

Minimum threshold $= 16 - 3.25 = 12.75$.

Therefore, according to maximum and minimum threshold values, a gateway with load 22 is considered a heavily loaded gateway while gateway with load 9 is considered an underloaded gateway. Finally, the fitness value calculated according to Equation 5.4 is *Fitness* $= (1 - (18/22)) + (2/4) = 0.6819$.

Let us consider an ideal case in Example 4.

Example 4: Consider g_1, g_2, g_3 and g_4 have loads 5, 5, 5 and 5, respectively.

From Equation 5.7, the mean of loads is $= \frac{(5+5+5+5)}{4} = 5$.

Range of gateway load using Equation 5.9 is $= 5 - 5 = 0$.

$\gamma = 0$ as range is zero so there are no heavy and underloaded gateways.

Finally, fitness $= (1\text{-}1) + 0 = 0$.

Therefore, fitness value zero indicates that the network is fully balanced. In real time scenario, it is not possible to have a fully balanced network; therefore it is better to choose a solution with fitness value closer to zero.

5.3.4　Formation of memeplexes phase

Each frog (Solution) is evaluated according to the fitness function. All frogs are sorted in the ascending order in an array X. Further the array X is partitioned into m memeplexes such that first frog is allocated to first memeplex, second frog allocated to the second memeplex, m^{th} frog allocated to the mth memeplex, and $(m+1)^{th}$ frog is allocated to the first memeplex, and so on. This entire procedure is explained according to Example 2, with the number of memeplexes equal to 2 in Figure 5.3 and 5.4. Figure 5.3 shows randomly generated eight frogs and Figure 5.4 shows it generated two memeplexes using eight frogs.

5.3.5　Formation of sub-memeplexes phase

In Figure 5.4 M_1 and M_2 are considered as two memeplexes. Now each memeplex is divided into number of sub-memeplexes. Number of sub-

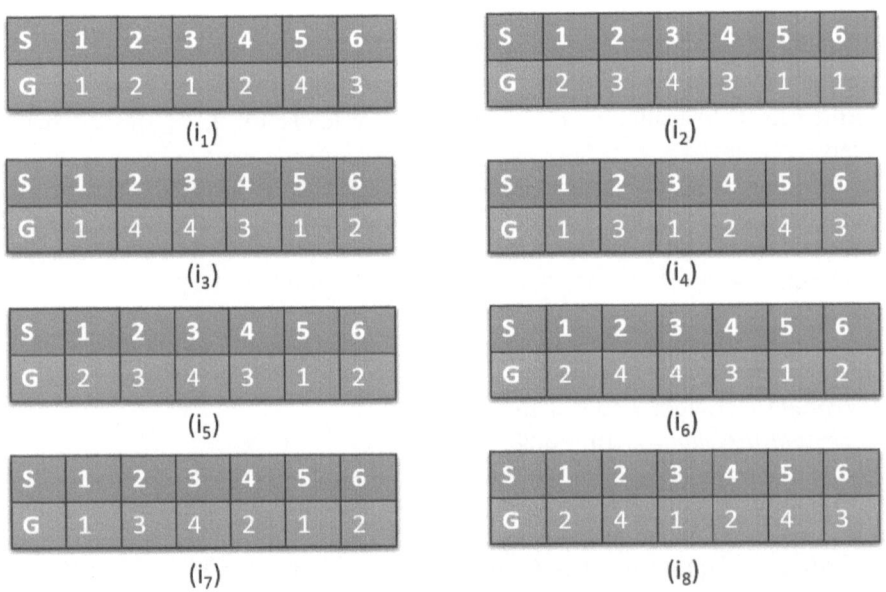

FIGURE 5.3: Eight solutions from initial population

FIGURE 5.4: Formation of 2 memeplexes from 8 frogs

	N_1	N_2
M_1	i_1, i_3	i_5, i_7
M_2	i_2, i_4	i_6, i_8

FIGURE 5.5: Formation of 4 sub-memeplexes from 2 memeplexes

memeplexes are determined by randomly selecting one of the factors of $\frac{population-size}{m}$. Each sub-memeplex is denoted as (M_i, N_j), where i and j represents memeplex and sub-memeplex numbers, respectively. The Figure 5.5 shows formation of 4 sub-memeplexes from 2 memeplexes.

5.3.6   Offspring generation phase

In the case of generating two offsprings from parents, there is a chance of losing the best parent. To resolve this issue, we are replacing only the worst parent with best offspring. The offspring is produced by exchanging the information from P_B (best solution in sub-memeplex) to P_W (worst solution in sub-memeplex). Replacing P_W with new offspring preserves the P_B. Generate a random point q such that $1 < q \leq k$, where k is number of sensor nodes. Copy all the gateway information after sensor node q from P_B to P_W. P_X is the first solution in each memeplex, and P_G is the global best solution.

By continuing the Example 2, in sub-memeplex (M_2, N_1) (see Figure 5.5) solution i_2 is considered as P_B and i_4 is considered as P_W. Here, we have used single point for information exchange as it gives better results than double and random point approaches [62]. This process is explained in Figure 5.6 using single point crossover for random point $q = 3$.

This process is carried out for all sub-memeplexes to generate offsprings. Next, the fitness value for all offsprings is evaluated.

Lemma: *The offspring produced by offspring generation phase is valid.*
Proof: A valid offspring is the one with every sensor node assigned to a gateway that is in its communication range. During the initial population generation phase, the gateway allocated to each sensor node is selected from the respective gateway list that is in its communication range. During information exchanging between P_B and P_W, the gateway associated with each sensor node in P_B and P_W is one of the gateways from its gateway list. Hence, for every sensor node, its corresponding gateway is valid.

Therefore, the offspring produced by offspring generation phase is valid.

5.3.7 Transfer phase

In this phase, a heavily loaded gateway is selected, and its load is reduced by transferring the farthest sensor node in its range to another gateway which is in range of the farthest sensor node. This helps in balancing the gateway's load and gateway's energy consumption (as we are transferring farther sensor node to another gateway). This phase is shown in Figure 5.7 for an offspring obtained in Figure 5.6.

According to Figure 5.6, gateway g_1 is heavily loaded (four sensor nodes are connected to it). During transfer phase, s_6 is transferred from g_1 to g_2 (assume s_6 is farther node from g_1). So, after transfer phase, new offspring is produced as shown in Figure 5.7.

Lemma: *The offspring produced by transfer phase is valid.*
Proof: While transferring the sensor node from a heavily loaded gateway to another gateway, we are selecting a gateway which is in communication range of respective sensor node. Hence, it produces a valid offspring.

Therefore, the offspring produced by transfer phase is valid.

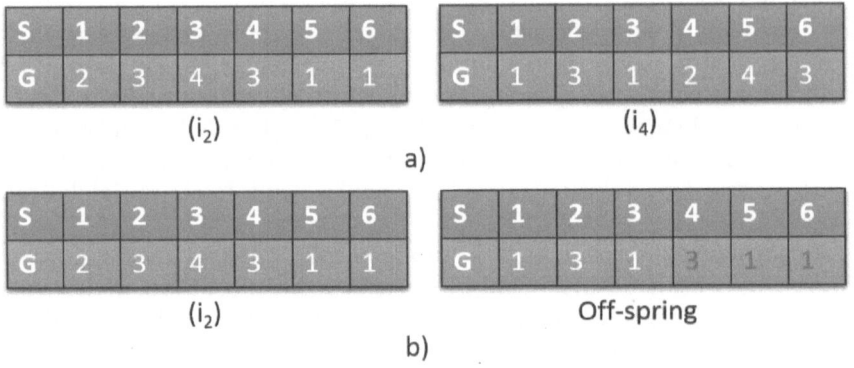

FIGURE 5.6: Offspring generation: a) Before; b) After

S	1	2	3	4	5	6
G	1	3	1	3	1	②

Off-spring

FIGURE 5.7: New offspring produced after transfer phase

5.3.8 Convergence checking phase

The fitness value for the new offsprings is calculated. These new offsprings are again sorted in ascending order in an array X. This causes shuffling of the frogs which are again partitioned, and evolution process continues until the convergence criteria is satisfied or up to the predefined number of iterations.

5.3.9 Algorithm description

Algorithm 2 shows the pseudo-code for the entire proposed ISFLA algorithm for load balancing in WSNs.

5.4 Results and discussion

5.4.1 Experimental setup

The experiments are performed for 50, 100 and 150 number of sensor nodes and 10, 15 and 20 number of gateways. The sensor nodes and gateways are deployed in a 50×50 m^2 area. The simulation parameters used for conducting experiments are mentioned in Table 5.2. Some of the settings need to be tuned finely to obtain the optimal solution. These parameters are also listed in Table 5.3.

For comparison purposes, along with the proposed ISFLA approach, we have implemented MSFLA [65], SFLA, Simple GA Load Balancing (SGA) [63], Novel GA Load Balancing (NGA) [19], NLDLB [14] and Score Based Load Balancing (SBLB) [15] algorithms in MATLAB. The evaluation factors considered for experimentation are energy consumption, heavy loaded sensor nodes and fitness, active and dead sensor nodes.

TABLE 5.2: Simulation parameters for ISFLA

Parameter	Value
Area	$50 \times 50 \ m^2$
Base Station Location	$(25, 25)$
Communication Range	$10m$
E_{elec}	$50 \ nJ/bit$
ϵ_{fs}	$10 \ pJ/bit/m^2$
ϵ_{mp}	$0.001 \ pJ/bit/ \ m^4$
Number of sensors	50, 100, 150
Number of gateways	10, 15, 20
Number of simulations	50

Algorithm 2 Pseudo code for ISFLA-based Load Balancing

1: $S \leftarrow \{s_1, \ldots, s_k\}$ // A set of k sensor nodes.
2: $G \leftarrow \{g_1, \ldots, g_t\}$, where $t < k$. // A set of t gateways.
3: $Com(s_i)$ // Set of gateways in s_i communication range.
4: $PopSize$ // Initial number of frogs.
5: $F \leftarrow \{i_1, i_2, \ldots, i_{PopSize}\}$. // A set of solutions.
6: P_B, P_W // Best and worst frog in sub-memeplex.
7: P_G // Global best frog among all memeplexes.
8: $M \leftarrow \text{fact}(PopSize)$ // Factors of $PopSize$.
9: $N \leftarrow \text{fact}(M)$.
10: β // Number of generations. An assignment A: S → G
11: **Step 1: Initialization phase**
12: **for** $j = 1$ to β **do**
13:     **for** $i = 1$ to k **do**
14:         $g_i \leftarrow \{s_i \mid g_i \in Com(s_i)\}$. // Randomly assign s_i to $g_i \in Com(s_i)$.
15:     **end for**
16: **Step 2: Fitness evaluation and sorting phase**
17:     Find fitness value for every frog in F by using Eqs. (5.4)-(5.10).
18:     Sort F with ascending order of frog's fitness value.
19: **Step 3: Formation of memeplexes**
20:     $X \leftarrow F$.
21:     $m \leftarrow \text{rand}(M)$.
22:     Partition F into m memeplexes as $M_1, M_2, \ldots, M_m \mid X_1 \in M_1, X_2 \in M_2, \ldots, X_m \in M_m, X_{m+1} \in M_1$ and $||M_i|| = k$, $forall$ i=1, $\ldots$, m.
23:     $P_G \leftarrow X_1$.
24: **Step 4: Formation of sub-memeplexes phase**
25:     **for** $i = 1$ to m **do**
26:         $n \leftarrow \text{rand}(m)$ // Select a point n randomly in M_i.
27:         Make n sub-memeplexes from M memeplexes.
28:     **end for**
29: **Step 5: Local exploration**
30:     **for each** sub-memeplex$\in \{(M_1, N_1), \ldots, (M_m, N_n)\}$ **do**
31:         $q \leftarrow \text{rand}(k)$.
32:         **Step 5.1: Offspring generation phase**
33:         $P_W(g_{q+1}, \ldots, g_k) \leftarrow P_B(g_{q+1}, \ldots, g_k)$ // Generate offspring by copying gateway information after point q from P_B to P_W.
34:         $G_x \leftarrow \text{mode}(P_W(G))$. // Find heavily loaded gateway in offspring.
35:         **Step 5.2: Transfer phase**
36:         Reduce the gateway load by transferring farthest sensor node (S_f) from G_x to another gateway $G_y \in Com(S_f)$.
37:     **end for each**
38: **Step 6: Convergence checking**
39:     Evaluate each new offspring produced in **Step 5**.
40:     **if** convergence criteria are satisfied **then**
41:         Stop
42:     **else**
43:         go to **Line no. 17**
44:     **end if**
45: **end for**

TABLE 5.3: Tuning parameter values for proposed SFLA initialization

Parameter	Value	Explanation
PopSize	50	Initial population size
α	10	Number of generations for offspring
β	500	Number of generations for solution
m	2 to 25	Number of memeplexes
n	2 to 5	Number of sub-memeplexes in each memeplex

5.4.2   Number of sensor nodes versus energy consumed

We have considered 50, 100 and 150 number of sensor nodes with an equal and unequal load on sensor nodes. The proposed approach is compared with MSFLA, SFLA, NLDLB, SBLB, SGA and NGA. The energy consumption until the first gateway dies is calculated in case of each algorithm. Figure 5.8 (a) and Figure 5.8 (b) show the results of energy consumption for an equal and unequal load of the sensor nodes, respectively. It is observed that the proposed ISFLA consumes less amount of energy than the energy consumption in other compared algorithms. This is because the solution selected by the SFLA approach is an energy-efficient solution according to the fitness function. Hence, at each generation, the energy is saved.

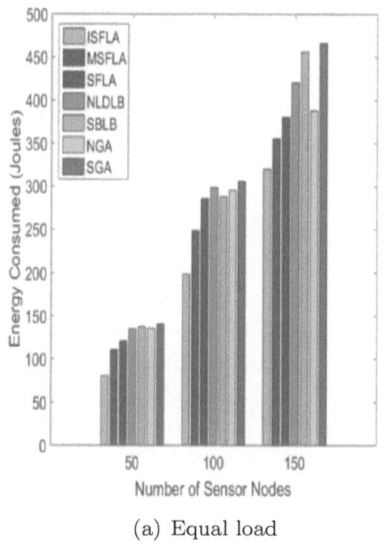

(a) Equal load

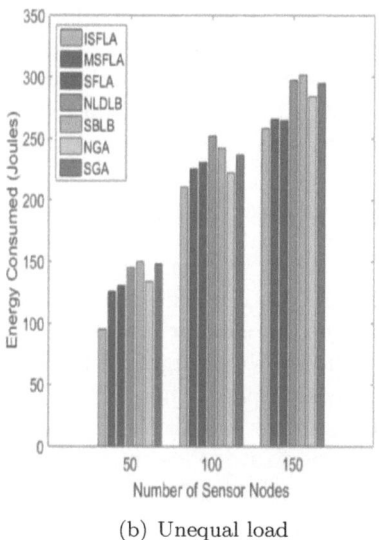

(b) Unequal load

FIGURE 5.8: Comparison of the proposed ISFLA algorithm with MSFLA, SFLA, NLDLB, SBLB, SGA and NGA algorithms in terms of number of sensor nodes and energy consumption for: (a) Equal load and (b) Unequal load

5.4.3　Number of heavy loaded sensor nodes versus fitness

We have performed experiments on the scenarios with 50, 100 and 150 number of sensor nodes by varying the load of one sensor node to five sensor nodes. The load variation means the different number of data packets sent by the sensor node at a time. To validate the performance of the proposed approach, we have implemented MSFLA, SFLA, NLDLB, SBLB, SGA, NGA algorithms. Figure 5.9 (a), Figure 5.9 (b) and Figure 5.9 (c) show the fitness values of the solutions produced by the scenarios with 50, 100 and 150 number of sensor nodes, respectively. It is observed that the proposed ISFLA approach achieves best fitness value even for heavily loaded sensor nodes. This is due to the designed fitness function which always considers the granted gateways in the solution, that is, the load-balanced solution.

5.4.4　Number of generations versus fitness

To evaluate this parameter, we have considered only meta-heuristic approaches, namely MSFLA, SFLA, SGA and NGA. NLDLB and SBLB algorithms are heuristic approaches, so they could not produce solutions in generations. The scenarios considered for experimentation are with 50, 100 and 150 number of sensor nodes. Figure 5.10 (a) and Figure 5.10 (b) show the results using 50 sensor nodes for equal and unequal load, respectively. Figure 5.10 (c) and Figure 5.10 (d) show the results using 100 sensor nodes for equal and unequal load, respectively. Figure 5.10 (e) and Figure 5.10 (f) show the results using 150 sensor nodes for equal and unequal load, respectively. Results from Figure 5.10 show that ISFLA performs better than other compared algorithms. Here, the fitness function is a minimizing fitness function; therefore, at each iteration, the solution is improved with minimum fitness value. Also, there is no loss of best solution at each iteration.

5.4.5　First node die

This parameter defines the time (round) at which the death of the first sensor node occurs. The experiments are carried out for 50, 100 and 150 number of sensor nodes for both equal and unequal load on the sensor nodes. The results are shown in Figure 5.11 (a) and (b) for equal and unequal load, respectively. It is observed that ISFLA outperforms NGALB, SGALB, NLDLB and SBLB algorithms with the higher round number. This is because the energy consumption rate is less in ISFLA than other algorithms.

5.4.6　Half of the nodes alive

This parameter denotes the round number, where half of the sensor nodes are active. The time difference between the death of the first sensor node and the death of half the sensor nodes defines the sustainability of the network. The results of this parameter are shown in Figure 5.12 (a) and (b) for an equal

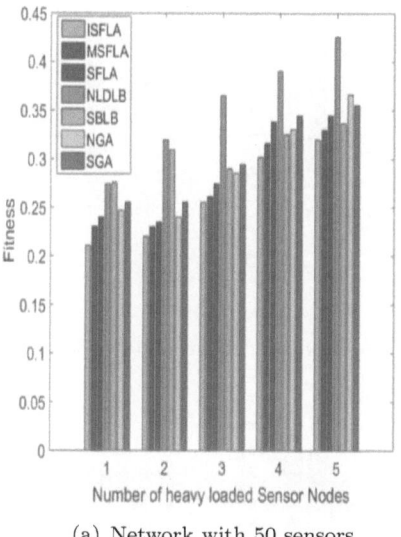

(a) Network with 50 sensors

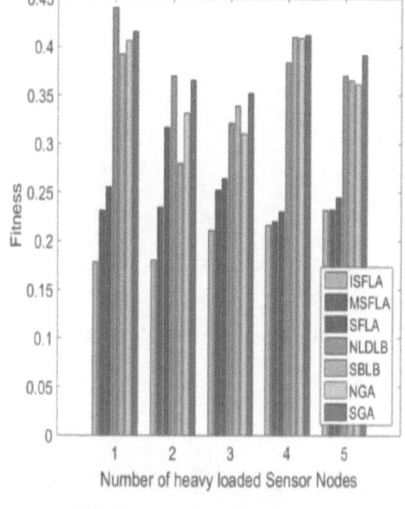

(b) Network with 100 sensors

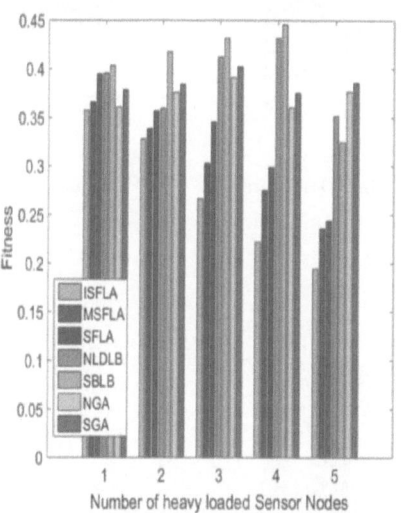

(c) Network with 150 sensors

FIGURE 5.9: Comparison of the proposed ISFLA algorithm with MSFLA, SFLA, NLDLB, SBLB, SGA and NGA algorithms in terms of number of heavy loaded sensor nodes and fitness value for: (a) Network with 50 sensor nodes, (b) Network with 100 sensor nodes and (c) Network with 150 sensor nodes

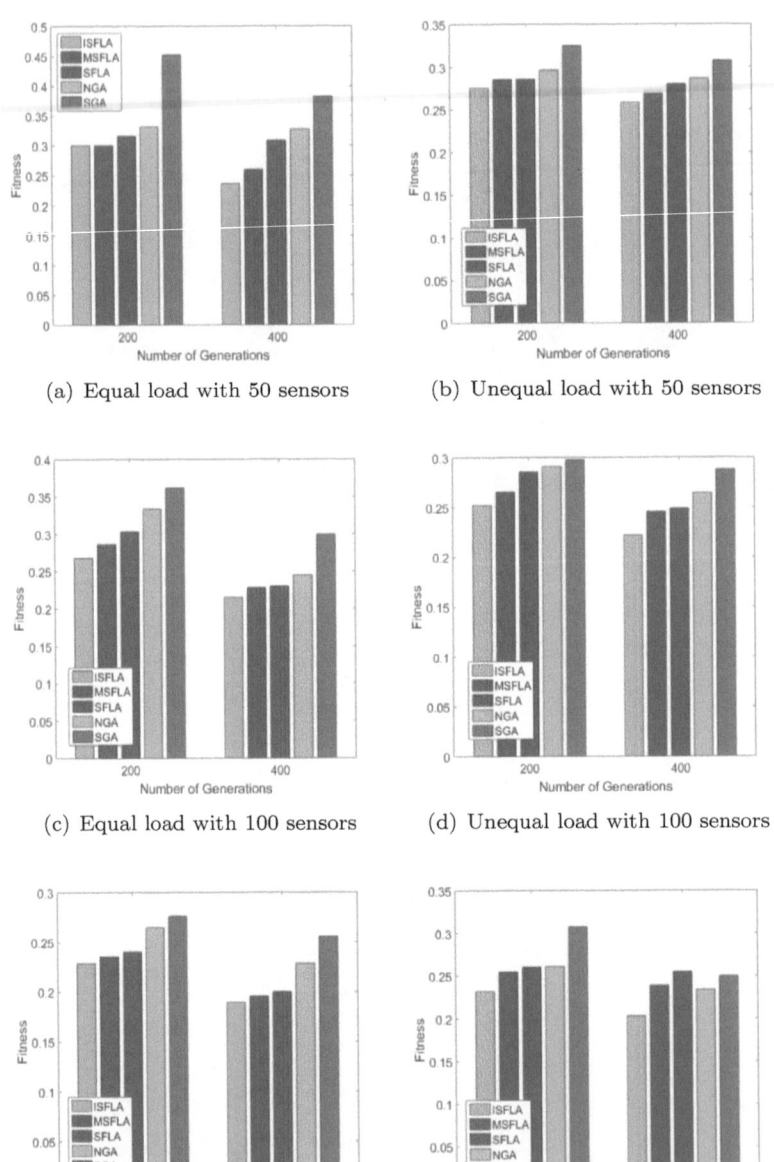

(a) Equal load with 50 sensors   (b) Unequal load with 50 sensors

(c) Equal load with 100 sensors   (d) Unequal load with 100 sensors

(e) Equal load with 150 sensors   (f) Unequal load with 150 sensors

FIGURE 5.10: Comparison of the proposed ISFLA with MSFLA, SFLA, SGA and NGA in terms of number of generations and fitness value for (a) Equal load with 50 sensor nodes, (b) Unequal load with 50 sensor nodes, (c) Equal load with 100 sensor nodes, (d) Unequal load with 100 sensor nodes, (e) Equal load with 150 sensor nodes and (f) Unequal load with 150 sensor nodes

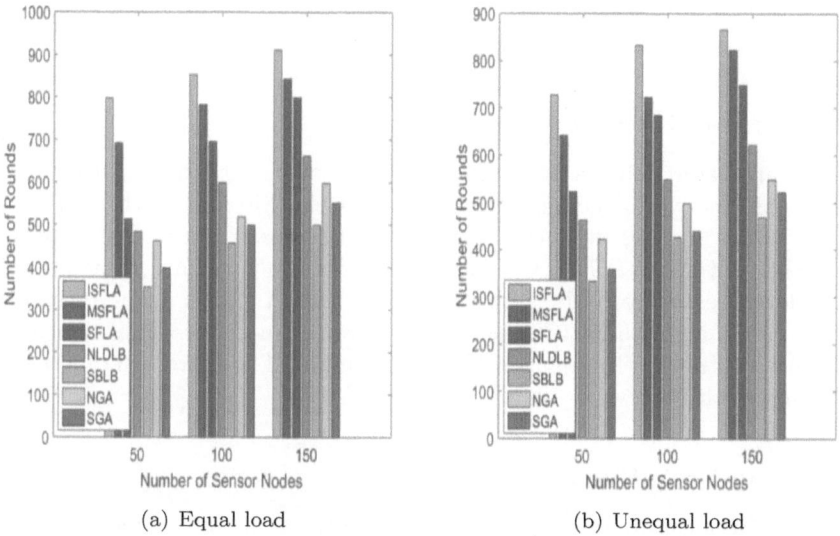

(a) Equal load   (b) Unequal load

FIGURE 5.11: Comparison of the proposed algorithm with NGALB, SGALB, NLDLB and SBLB algorithms in terms of first node dies for: (a) Equal load; (b) Unequal load

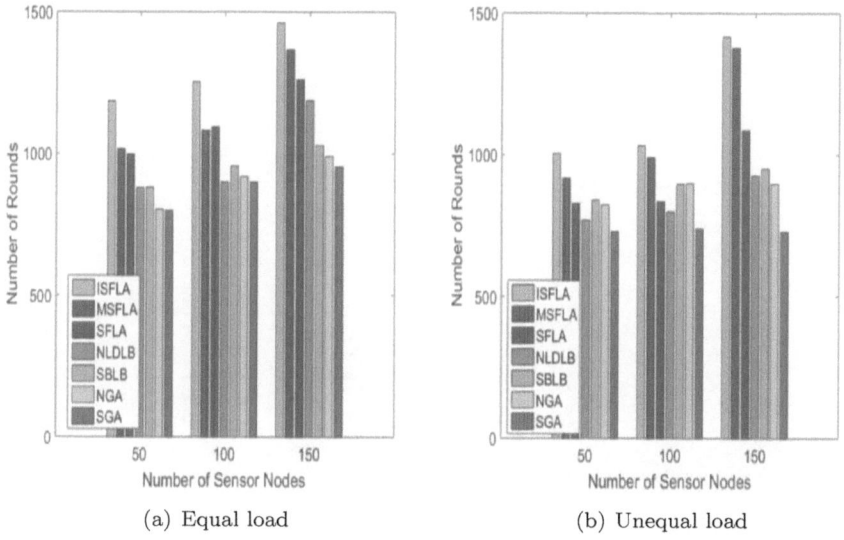

(a) Equal load   (b) Unequal load

FIGURE 5.12: Comparison of the proposed algorithm with NGALB, SGALB, NLDLB and SBLB algorithms in terms of half of the nodes die for: (a) Equal load; (b) Unequal load

and unequal load of the 50, 100 and 150 number of sensor nodes. Proposed ISFLA shows outperformance in both the cases and all of the three scenarios. The network generated by ISFLA survives for a longer time.

5.4.7 First gateway death

This parameter defines the round at which the death of the first gateway from the network occurs. This represents the lifetime of the network. Figure 5.13 (a) and (b) denotes the results of the death of the first gateway from ISFLA, NGALB, SGALB, NLDLB and SBLB algorithms for equal and unequal load, respectively. It is observed that ISFLA shows outperformance over compared algorithms with the highest round number. The algorithm selects the solution with balanced load constraints. Therefore, this increases the lifetime of the gateway.

5.4.8 Number of dead sensor nodes

The experiments are performed for the scenario with 150 number of sensor nodes. The round at which the death of 10, 20, 30, 40 and 50 number of dead sensor nodes is noted. Figure 5.14 (a) and (b) shows the performance of the ISFLA, NGALB, SGALB, NLDLB and SBLB algorithms under this

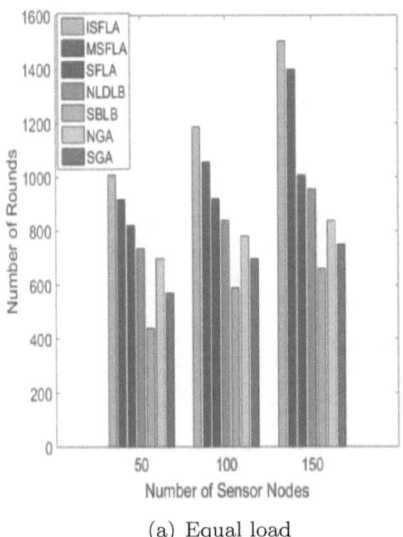

(a) Equal load

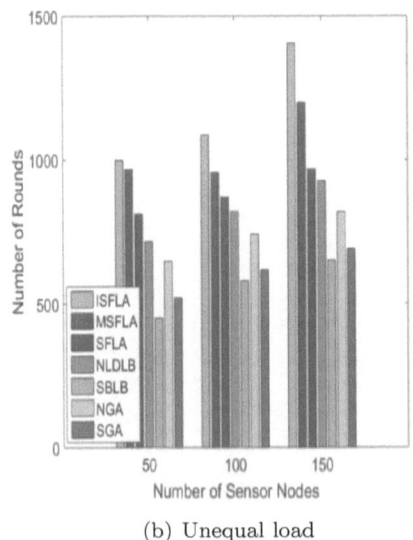

(b) Unequal load

FIGURE 5.13: Comparison of the proposed algorithm with NGALB, SGALB, NLDLB and SBLB algorithms in terms of first gateway die for: (a) Equal load; (b) Unequal load

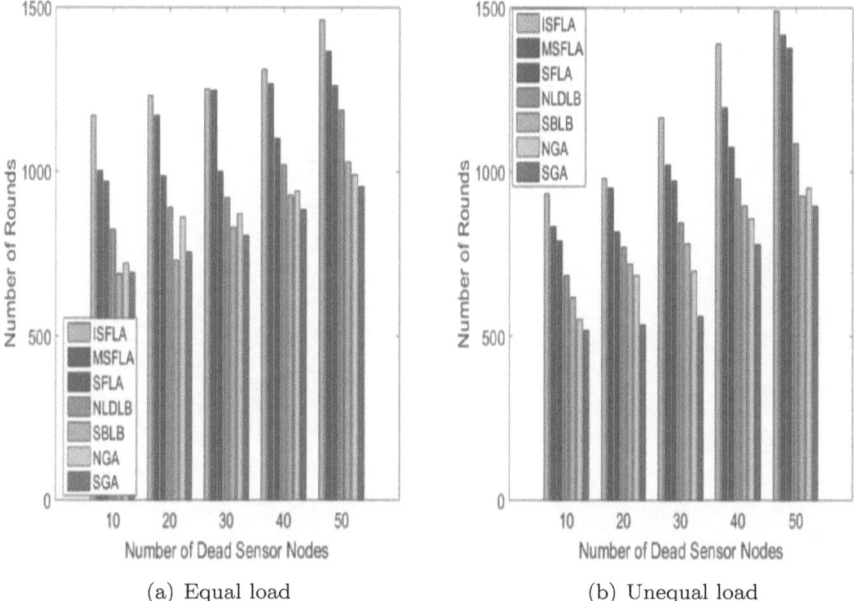

(a) Equal load   (b) Unequal load

FIGURE 5.14: Comparison of the proposed algorithm with NGALB, SGALB, NLDLB and SBLB algorithms in terms of dead sensor nodes for: (a) Equal load; (b) Unequal load

parameter for equal and unequal load, respectively. It is found that, in ISFLA, the rate of the death of the sensor node is less than other algorithms. This is due to the energy consumption rate being less in ISFLA than other algorithms.

5.5 Conclusion

SFLA is improved according to the application of WSN. The improvements have been made in the phases of the initial population, offspring generation. A novel fitness function is designed by considering the residual energy of gateways. To prove the effectiveness of the proposed approach, it is compared with the state-of-the-art algorithms. It is observed that proposed ISFLA outperforms these algorithms under various evaluation factors.

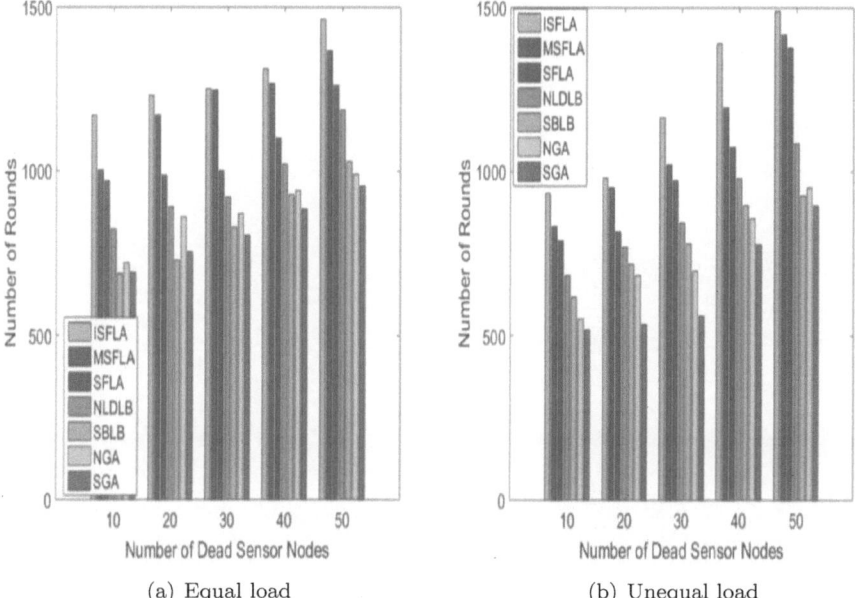

(a) Equal load   (b) Unequal load

FIGURE 5.14: Comparison of the proposed algorithm with NGALB, SGALB, NLDLB and SBLB algorithms in terms of dead sensor nodes for: (a) Equal load; (b) Unequal load

parameter for equal and unequal load, respectively. It is found that, in ISFLA, the rate of the death of the sensor node is less than other algorithms. This is due to the energy consumption rate being less in ISFLA than other algorithms.

5.5 Conclusion

SFLA is improved according to the application of WSN. The improvements have been made in the phases of the initial population, offspring generation. A novel fitness function is designed by considering the residual energy of gateways. To prove the effectiveness of the proposed approach, it is compared with the state-of-the-art algorithms. It is observed that proposed ISFLA outperforms these algorithms under various evaluation factors.

Chapter 6

SCE-PSO Based Clustering Technique for Load Balancing in WSN

6.1 Introduction

In recent years, the advancement of soft computing has given more focus on bio-inspired algorithms. Many researchers have implemented various algorithms to solve different problems of WSNs. In clustered WSNs, load balancing of CHs is a crucial factor to operate the network for longer time. Therefore, an efficient clustering technique can solve the problem of load balancing. In this chapter, we present an SCE-PSO based clustering technique with a novel fitness function to enhance the lifetime of the network.

6.2 Preliminaries

6.2.1 Terminologies

The following terminology has been used in the proposed SCE-PSO algorithm and a novel fitness function.

1. The set of sensor nodes is represented by $S = \{s_1, s_2, \ldots, s_m\}$.// where m = Number of sensor nodes

2. The set of gateways is represented by $G = \{g_1, g_2, \ldots, g_n\}$. // where n = Number of gateways

3. $\text{Dist}(s_i, g_j)$: The distance between the sensor node s_i and the gateway g_j.

4. $Range(s_i)$: The total number of gateways within contact range of s_i.

5. Load(g_i): It depends on the amount of energy needed for gateway g_i to receive and transmit the l bits of data ($E(l)$) and the residual energy of gateway.

$$\text{Load}(g_i) = E(l)/\text{residual energy of } g_i \qquad (6.1)$$

where $E(l)$ is computed based on the energy model and it is given below. $E(l) = E_R(l) + E_T(l, d)$

6. Load Factor(g_i): It is the ratio between the Load(g_i) and maximum load of the gateway in the network.

$$\text{Load Factor}(g_i) = \frac{\text{Load}(g_i)}{\max\{\text{Load}(g_i), \forall i = 1, 2, \ldots n\}} \qquad (6.2)$$

7. Expected Gateway Factor (EGF): It is the ratio between the sum of load factors of all gateways and total number of gateways in the network.

$$\text{EGF} = \left[\frac{\sum_{i=1}^{n} \text{Load Factor}(g_i)}{\text{Total number of gateways}} \right] \qquad (6.3)$$

8. MeanCluDist(g_j): It denotes the average Euclidean distance of each member sensor node to gateway g_j.

$$\text{MeanCluDist}(g_j) = \left[\frac{\sum_{i=1}^{m} \text{Dist}(s_i, g_j) * \alpha_{ij}}{K_j} \right] \qquad (6.4)$$

where K_j denotes total number of sensor nodes associated to gateway g_j, α_{ij} is a boolean variable, $\alpha_{ij} = 1$ if the sensor node s_i is a member sensor node of gateway g_j, otherwise, $\alpha_{ij} = 0$.

6.3   Overview of SCE-PSO

6.3.1   Background of SCE-PSO

The SCE-PSO is an evolutionary algorithm, implemented by [87]. The PSO is a population-based meta-heuristic approach inspired by birds [88], [89]. A population of points (sample size) in the SCE-PSO is randomly sampled from the available space. SCE-PSO divides the population into a specified number of complexes based on the fitness value, and it evaluates each complex using the PSO algorithm. The regular PSO computes every particle in the population, although SCE-PSO selects a sub-swarm from the population for evaluation. The SCE-PSO mainly consists of six phases: (1) Initialization of random population, (2) Ranking the solutions or points, (3) Divide the solutions into complexes, (4) Complex evolution using PSO, (5) Shuffling of complexes and (6) Convergence checking. The detailed description of the SCE-PSO algorithm is summarized below, and the flow chart for the SCE-PSO algorithm is represented in Figure 6.1.

Step 1. Initialization:
Let $N \geq 1$, $M \geq 1$ be two numbers, where N denotes the number of complexes, M denotes the number of particle in each complex. Compute the size of the sample $S = NM$ and take $P_1, P_2, \ldots, P_S$ to be S random sample points (particles). Initialize each particle in the sample and evaluate fitness value f_i at each particle P_i.

Step 2. Ranking:
Sort the sample points or particles in an array A based on the fitness value.
A= $\{P_i, f_i, \text{i=1, 2, 3,} \ldots, \text{S}\}$

Step 3. Partitioning:
Divide an array A into N number of complexes, i.e., $C_1, C_2, \ldots, C_N$. Each complex consists of M particles in such a way that the particle P_1 is assigned to complex C_1, the particle P_2 assigned to complex C_2, the particle N is assigned to the C_N complex, and the particle $N+1$ assigned to the C_1 complex, and so on.

Divide an array A into N complexes $\{C_1, C_2, \ldots, C_N\}$, and each complex consist of M points,
such that: $C_x = \{P_x^j, f_x^j | P_x^j = P_{x+N(j-1)}, f_x^j = f_{x+N(j-1), j=1,2,\ldots,M}\}$

Step 4. Complex Evolution:
Each complex C_x $\{x = 1, 2, \ldots, N\}$ is being evaluated according to PSO algorithm. PSO algorithm is described further. Regular PSO computes each particle in the population, although SCE-PSO selects sub-swarm from the complex for assessment based on the fitness value. Let us consider that the sub-swarm contains the Q number of points and apply PSO algorithm for selected Q points.

Step 5. Complex Shuffling:
Restore $C_1, C_2, \ldots, C_N$ into an array A. Sort an array A according to its fitness value.

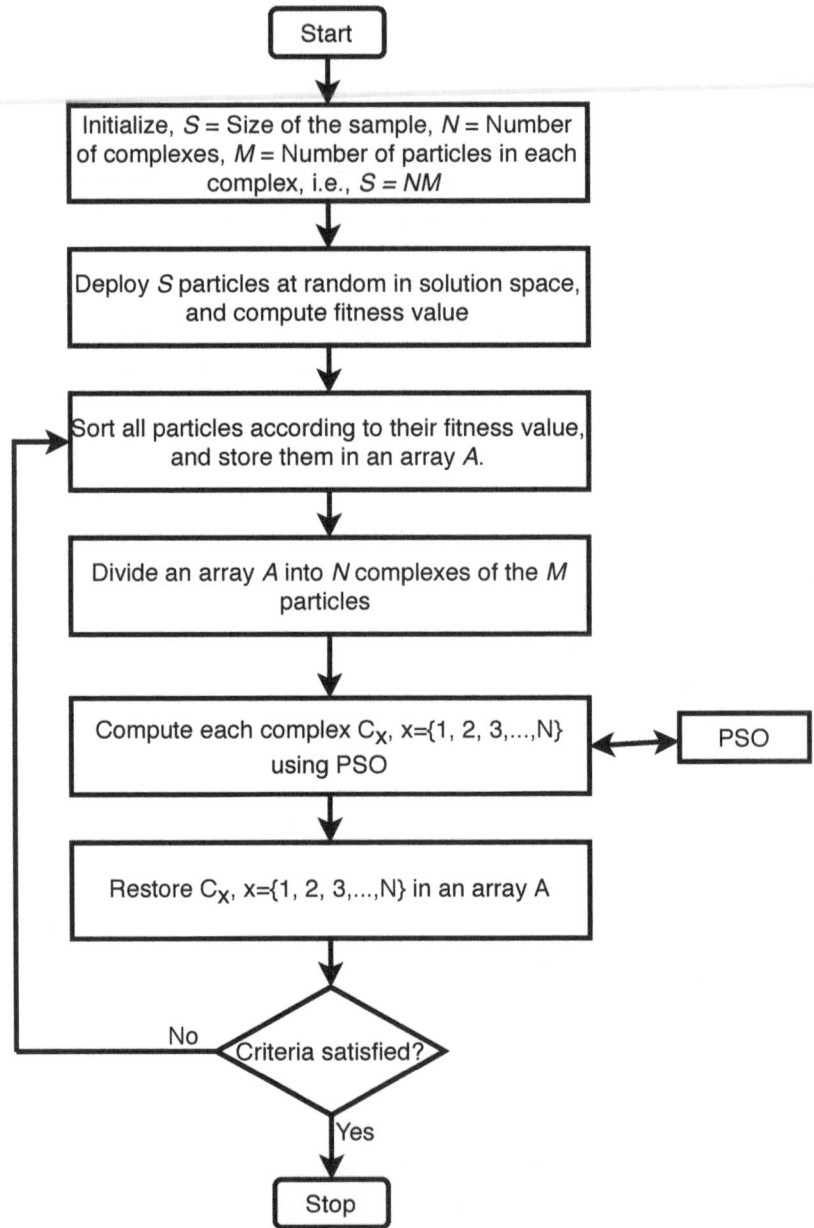

FIGURE 6.1: Flowchart of SCE-PSO algorithm

Step 6. **Convergence Checking:**
Check whether convergence conditions are met or not. If convergence criteria are satisfied, stop; otherwise go to Step 3.

6.3.2   Background of PSO

PSO is inspired by nature based on bird flocking [88], [89]. Birds regularly fly together in a group in search of food or shelter, without colliding. Each member or bird in a group follows the group's information by changing its velocity and location. Any effort made by each bird or individual member to find food and shelter in a group is reduced by sharing group knowledge. Every particle in a complex gives a solution to a particular problem case, and a fitness function can validate the quality of each particle. All particles have equal dimensions. Each particle P_i has a position (X_{id}) and velocity (V_{id}) in d^{th} dimension of the hyperspace. So, at any point in time, particle P_i is interpreted as

$$P_i = \{X_{i1}, X_{i2}, X_{i3}, \ldots, X_{id}\}$$

To get an efficient solution, each particle P_i follows its own best position called *Lbest$_i$* and global best called *Gbest* to update its position and velocity recursively. The *Gbest* is the best solution among all particles in the population. The below equations represent the velocity (V_{id}) and position (X_{id}) updating of particle P_i.

$$V_{id}(t+1) = W * V_{id}(t) + a_1 * r_1 * (Lbest_{id} - X_{id}(t))$$
$$+ a_2 * r_2 * (Gbest_d - X_{id}(t)) \tag{6.5}$$

$$X_{id}(t+1) = X_{id}(t) + V_{id}(t+1) \tag{6.6}$$

where W indicates the inertial weight, a_1 and a_2 represent the acceleration constants that are non-negative real numbers. r_1 and r_2 are random numbers within the range of [0,1] that are evenly distributed. Repeat the update process until we identify the required solution or hit the maximum number of iterations. The complete description of the flow is represented in Figure 6.2. Using Equation 6.5 and Equation 6.6, the position and the velocity of each particle is modified at each iteration. Therefore, if the position of the particle goes out of search space, then change it to the previous value.

6.4   Proposed SCE-PSO based clustering

SCE-PSO strategy is used to solve the problem of load balancing of gateways in the WSN, and it is an optimization algorithm. The SCE-PSO technique mainly contains six phases: (1) Random generation of particles in the population, (2) Fitness function evaluation, (3) Particle sorting and partition, (4) Complex evolution using PSO, (5) Complexes are shuffling and (6) Convergence checking. The complete description of the phases, as mentioned above, is summarized as follows.

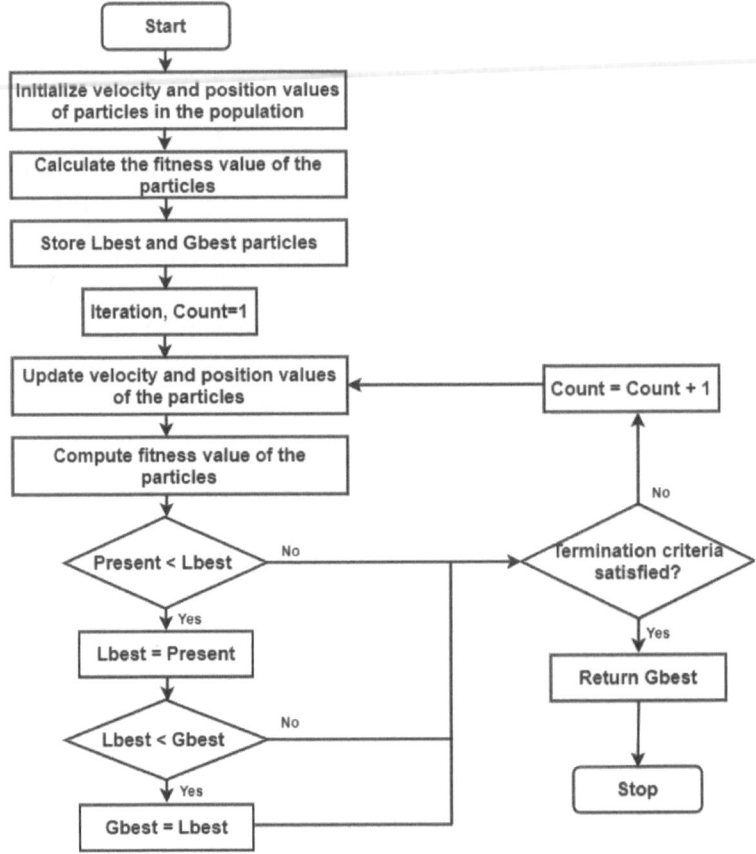

FIGURE 6.2: Flowchart of PSO algorithm

6.4.1   Random particle generation

Each particle in the sample is encoded as a solution to the clustering problem [90]. The dimension of each particle is D, and it is equal to all particles in the sample. In this problem, the D value of the particle is the number of sensor nodes in the network. Let $(X_{id}|1 \leq i \leq M, 1 \leq d \leq D)$ be the position value of the dth sensor node in the ith particle. Initialize each sensor node position in the particle with the randomly produced number from a uniformly distributed range of $(0, 1]$. The position value of dth sensor node called X_{id} assigns a gateway g_k (Let us consider k^{th} gateway) to s_d, i.e., sensor node s_d sends data to the g_k, and this can be described as follows:

$$n = ceiling(|X_{id} * Range(s_d)|) \tag{6.7}$$

i.e., g_k is the nth gateway in $Range(s_d)$. The generation of the random particle process is described in the following examples.

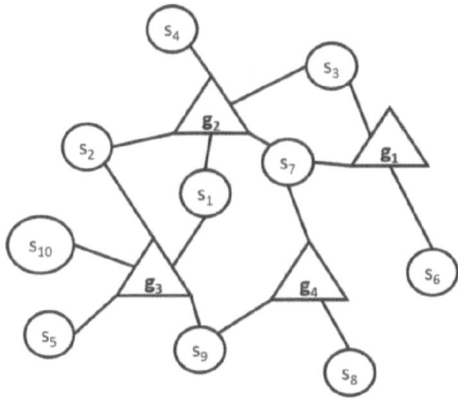

FIGURE 6.3: Random WSN scenario before clustering

Example 1: Let us consider a random WSN scenario with 4 gateways and 10 sensor nodes, i.e., $G = \{g_1, \ldots, g_4\}$ and $S = \{s_1, s_2, \ldots, s_{10}\}$. Therefore, in this scenario, the dimension D of the particle is 10. Figure 6.3 shows that the sensor node and their contact range and the link between s_i and g_j, and it shows the sensor node s_i is eligible to transfer data to the gateway g_j. The same process is represented in Table 6.1 in terms of sensor nodes and their contact range gateways.

During initialization, each dimension (sensor node) of X_{id} of particle P_i is initialized with a random number in (0,1] range. The allocated values for the random particle P_i are shown in Table 6.2. This random initialization of the location values of the sensor nodes is a solution for the problem, and it is generated as follows: Let us find the sensor node s_2, and its allotted random position value is 0.65. From the Equation 6.7, $ceiling(0.65 * 2) = 2$, i.e., the gateway (g_3) in the $Range(s_2)$ is allotted to sensor node s_2 for transmission of sensed data. The process continues for all the sensor nodes in the WSN scenario to get the solution to the problem. Table 6.2 represents the final result of network topology after clustering process, and the same is shown in Figure 6.4.

An initial sample is a group of particles that are produced at random. Each particle is the solution to the problem of load balancing. All particles are assumed to be valid if the sensor node is allocated to one of the gateways within the contact range of that sensor node. The following example explains this initial generation of the sample:

Example 2: From Figure 6.3, the WSN scenario contains 4 gateways and 10 sensor nodes. The scenario consists of 10 sensor nodes; the dimension of

TABLE 6.1: Sensor nodes and their contact range gateways

Sensor nodes	$Range(s_i)$	$|Range(s_i)|$
s_1	$\{\,g_2, g_3\,\}$	2
s_2	$\{\,g_2, g_3\,\}$	2
s_3	$\{\,g_1, g_2\,\}$	2
s_4	$\{\,g_2\,\}$	1
s_5	$\{\,g_3\,\}$	1
s_6	$\{\,g_1\,\}$	1
s_7	$\{\,g_1, g_2, g_4\,\}$	3
s_8	$\{\,g_4\,\}$	1
s_9	$\{\,g_3, g_4\,\}$	2
s_{10}	$\{\,g_3\,\}$	1

TABLE 6.2: Particle generation

Sensor nodes	X_{id}	$|Range(s_d)|$	$ceiling(|X_{id} * Range(s_d)|)$	Selected gateway
s_1	0.43	2	1	g_2
s_2	0.65	2	2	g_3
s_3	0.28	2	1	g_1
s_4	0.32	1	1	g_2
s_5	0.68	1	1	g_3
s_6	0.98	1	1	g_1
s_7	0.50	3	2	g_2
s_8	0.72	1	1	g_4
s_9	0.57	2	2	g_4
s_{10}	0.16	1	1	g_3

the randomly generated particle is 10. From Table 6.1, the sensor node s_1 can connect to either gateway g_2 or gateway g_3. In this example, the sensor node s_1 is chosen g_3 among g_2 and g_3; likewise s_2 selects gateway g_3 among g_2, g_3. s_7 selects g_2 among g_1, g_2, g_4, and so on. The above explained process generates the specified initial sample.

6.4.2 Evaluation of fitness function

The construction of the fitness function is a significant factor in WSN's design. The fitness function acknowledges the quality of the solution generated by the SCE-PSO. In this chapter, an efficient novel fitness function is designed to evaluate each particle in the sample. The novel fitness considers the expected gateway load, massive loaded gateways and mean cluster distance. The Equation 6.8 represents the fitness function to validate the particles in the sample.

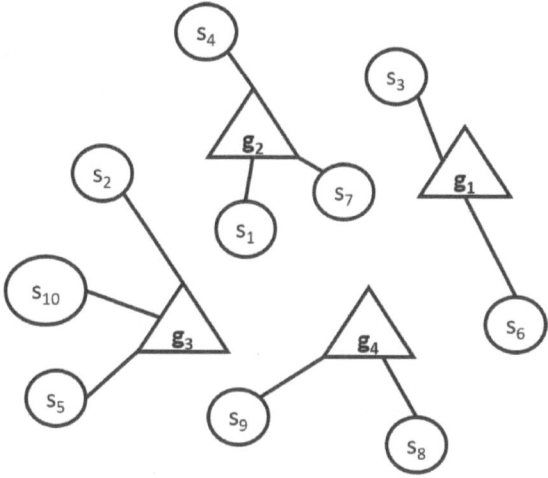

FIGURE 6.4: WSN after clustering

$$Fitness = \text{EGF} * \frac{\text{Total number of gateways}}{\text{Number of heavily loaded gateways}} * \frac{1}{\sigma_c} \quad (6.8)$$

Each gateway in the network is verified for whether the gateway is massively loaded or not in order to get the number of heavily loaded gateways on the network. The threshold value is formulated in Equation 6.9 to check the status of the gateway. The gateway whose load is more than the threshold value is called the heavily loaded gateway in the network.

$$Threshold = \frac{\sum_{i=1}^{n} \text{Load}(g_i)}{n} \quad (6.9)$$

$$Standard\ Deviation\ (\sigma_c) = \sqrt{\frac{\sum_{i=1}^{n} (\mu - MeanCluDist(g_i))^2}{n}} \quad (6.10)$$

$$\mu = \frac{\sum_{i=1}^{n} MeanCluDist(g_i)}{n} \quad (6.11)$$

where n = Number of gateways in the network.

It is observed from Equation 6.8 that the proposed fitness function acknowledges the gateway load, the mean cluster distance and the number of heavily loaded gateways to assess the particle quality. In Clustered WSN, sensor nodes transfer data to CHs or gateways. The number of connected sensor

TABLE 6.3: Partitioning of particles into complexes

C_1	P_1	P_4	P_7
C_2	P_2	P_5	P_8
C_3	P_3	P_6	P_9

nodes and distance to the sink increases the load of the gateway. Therefore, the number of allotted sensor nodes and distance between sink and gateway can decide the energy consumption rate of the gateway. Hence, balancing the load of the massively loaded gateway enhances the lifetime of the network. In this work, the major inclusion factor is reducing the energy usage of a massively loaded gateway and also considered is the mean cluster distance to reduce the energy usage of sensor nodes. Hence, the proposed fitness function is suitable for load balancing as well as energy efficiency. In the sample, the particle that is having a higher fitness value is a reliable solution for the clustering problem.

6.4.3 Particle sorting and partitioning

As mentioned above, each particle in the sample is computed by the fitness function to obtain the fitness value. Now, keep the particles in an array A as per the fitness values, where $A = \{P_i, f_i | i = 1, 2, \ldots, S\}$; here the first component of an array A represents a particle with the highest fitness value. Divide an array A into N complexes $C_1, C_2, \ldots, C_N$. Each complex consists of M particles in such a way that the particle P_1 is assigned to complex C_1, the particle P_2 assigned to complex C_2, the particle N is allocated to the C_N complex, and the particle $N + 1$ allocated to the C_1 complex, and so on, i.e.,

$$C_x = \{P_x^j, f_x^j | P_x^j = P_{x+N(j-1)}, f_x^j = f_{x+N(j-1), j=1,2,\ldots,M}\}$$

Example 3: Let us consider an approach consisting of 9 particles in the sample and 3 complexes, and each complex contains three particles. The 9 particles are captured randomly for the generation of the initial population. Consider a set of particles in the sample $S = \{P_1, P_2, \ldots, P_9\}$. These 9 particles are calculated using fitness function and stored in an array A according to the sorted order such that 1st component in A contains the highest fitness value particle and 9th component in A contains the lowest fitness value particle. Now, divide an array A into three complexes. Table 6.3 represents the complex partition of the sample.

6.4.4 Complex evolution

Each complex $C_x \{x = 1, 2, \ldots, N\}$ is being evaluated according to PSO algorithm. The SCE-PSO selects sub-swarm from the complex for validation

TABLE 6.4: Simulation parameters

Parameter	Value	Parameter	Value
parameters of WSN		Parameter of SCE-PSO	
Area	$50 * 50 \ m^2$	Maximum iterations	200
Sink	$(25, 25)$	a_1	1.4962
Communication Range	$10m$	a_2	1.4962
E_{elec}	$50 \ nJ/bit$	W	0.7968
ϵ_{fs}	$10 \ pJ/bit/m^2$	V_{min}	-0.5
ϵ_{mp}	$0.001 \ pJ/bit/ \ m^4$	Number of complexes	5 - 15
Sensor nodes	50, 100, 150		
Gateways	5, 11, 19		

based on the fitness value. Let us consider the sub-swarm that contains the Q number of points and apply the PSO algorithm for selected Q points. Now, run the PSO specified number of iterations using velocity and position using Equation 6.5 and Equation 6.6 in the sub-swarm. After updating velocity and positions in dimension D of a particle, we may get a better particle. The process of computing each complex using PSO is shown in Figure 6.2.

6.4.5   Complexes shuffling

Replace an array A with the updated or newly produced particles in the complexes $C_1, C_2, \ldots, C_N$. Sort an array A with descending order of their fitness value along with their position values.

6.4.6   Convergence checking

Check whether convergence conditions are met or not. If convergence criteria are satisfied, stop; otherwise continue the process until we meet requirements.

The complete process of the proposed SCE-PSO based clustering algorithm is described in Algorithm 3.

6.5   Results and discussion

Experiments are conducted by assuming a WSN scenario with a number of sensor nodes and gateways located at $50 * 50$ m^2 with a sensor node range of 10 meters. The simulation parameters for WSN and SCE-PSO algorithm are represented in Table 6.4.

Algorithm 3 Pseudo code for SCE-PSO based Clustering

- **Input**

  - Sensor node set $S = \{s_1, s_2, \ldots, s_m\}$.
  - Gateway set $G = \{g_1, g_2, \ldots, g_n\}$, where $n < m$.
  - $T = $ Total number of iterations.
  - $N = $ Total number of complexes.
  - $M = $ Number of particles in each complex.
  - $Q = $ Number of particles in each sub-swarm.

- **Output**

  - An assignment A: s $\rightarrow$ g

- **Algorithm**
  1: **phase 1: Initialization**
  2: - Initialize each particle position and velocity in the sample.
  3: **phase 2: Fitness Calculation**
  4: - Compute the fitness value of the particles.
  5: **phase 3: Sorting and partition**
  6: - Store all the particles in an array A with decreasing order of their fitness values.
  7: - Divide an array A into N complexes say $C_1, C_2, \ldots, C_N$ each containing M particles, such that $P_1 \in C_1, P_2 \in C_2, \ldots, P_N$ in $C_N, \text{N+1} \in C_1$.
  8: **phase 4: Complex Evolution**
  9: - Select Q particles from the complex.
  10: - Compute each complex using PSO.
  11: **phase 5: Complexes shuffling**
  12: - Replace $C_1, C_2, \ldots, C_N$ into an array A. Sort an array A with descending order according to their fitness value.
  13: **phase 6: Convergence checking**
  14: **if** convergence criteria are satisfied **then**
  15:     Stop
  16: **else**
  17:     goto **phase 3**
  18: **end if**

6.5.1 Performance analysis

The performance of the SCE-PSO algorithm is compared with the existing load balancing techniques such as NGA, SBLB, SGA and NLDLB. The efficiency of the SCE-PSO is validated in terms of network lifetime, total energy utilization and the number of gateways that dissolve their energy. The below subsections describe the measures as mentioned above in detail.

6.5.1.1 Network lifetime vs number of sensor nodes

Network lifetime is the total amount of time the node can operate functionally in the network, and it is measured in terms of rounds. The measure network lifetime depends on the application requirements and network structure. Experiments are conducted using $50 - 150$ sensor nodes with the equal and unequal load on sensor nodes. For comparison, tests are performed with NGA, SBLB, SGA and NLDLB and the SCE-PSO. The results of the simulations are shown in Figure 6.5 (a) and Figure 6.5 (b) for equal and unequal loads, respectively. It is noticeable from the figures that the SCE-PSO enhances the lifetime of the network as compared to NGA, SBLB, SGA and NLDLB. The SCE-PSO acknowledges the mean cluster distance, maximum loaded gateway and the number of heavily loaded gateways in the fitness.

6.5.1.2 Total energy utilization vs number of sensor nodes

Energy utilization is a measure of the energy consumed (in Joules) per round in the network. For comparison, we have conducted the experiments for $50 - 150$ sensor nodes with an equal and unequal load. The results of this analysis are in Figure 6.6. Results for equal and unequal loads represented are shown in Figure 6.6(a) and Figure 6.6(b), respectively. The figures show the total energy utilization across the network.

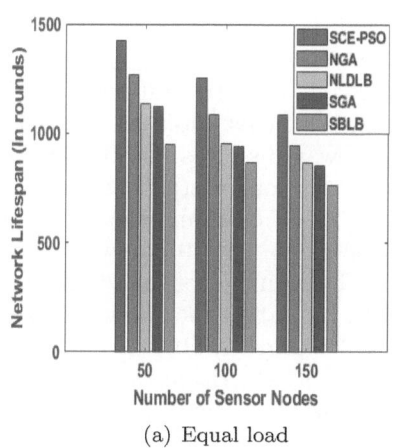

(a) Equal load

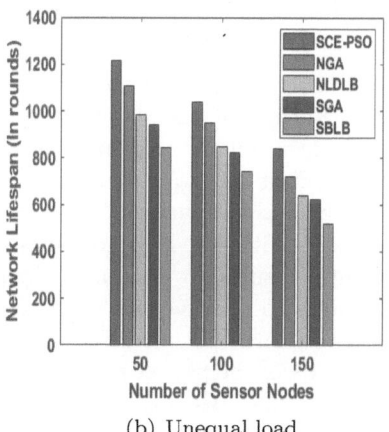

(b) Unequal load

FIGURE 6.5: Observation of the SCE-PSO with NGA, SBLB, SGA and NLDLB in terms of number of sensor nodes and network lifetime for: (a) Equal load and (b) Unequal load

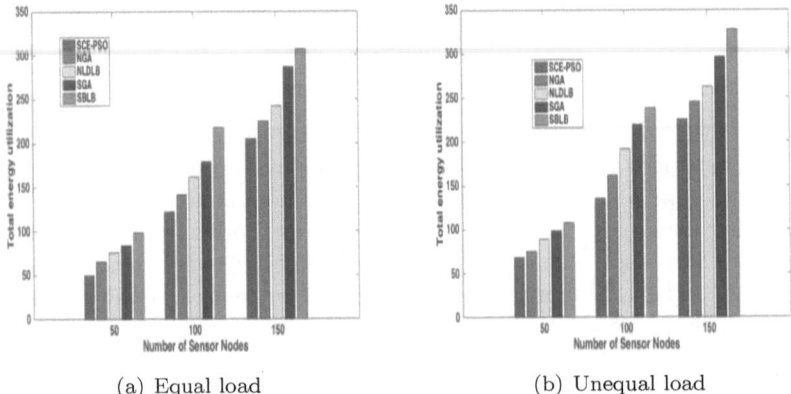

(a) Equal load   (b) Unequal load

FIGURE 6.6: Observation of the SCE-PSO with NGA, SBLB, SGA and NLDLB in terms of number of sensor nodes and total energy utilization for: (a) Equal load and (b) Unequal load

It is found from the results of the figures that the SCE-PSO algorithm provides significant results as compared to NGA, SBLB, SGA and NLDLB algorithms for both equal and unequal loads. The proposed fitness acknowledges the remaining energy of the gateways, anticipated load of the gateways when assigning the sensors and mean cluster distance. Thus, the solution provided by the SCE-PSO algorithm is efficient for load balancing problem.

6.5.1.3   First gateway that dissolves its energy and half of the gateways die

The first gateway that dissolves its energy is a metric estimate of the minimum lifetime of the gateway in the network. That indicates the number of rounds when the minimum lifespan gateway dissolves its total energy. The highest first gateway dies value indicates the longer stability period of the network. For comparison, NGA, SBLB, SGA, NLDLB and the SCE-PSO experimented with 50, 100, 150 sensor nodes and 5, 11, 19 gateways. Such experimental results are shown in Figure 6.7. The number of rounds the first gateway takes to dissolves its total energy and for half of the gateways to die in the network are shown in Figure 6.7 (a) and Figure 6.7 (b), respectively.

The results of the figures show that the SCE-PSO algorithm has the highest first gateway that dissolves its energy and half the gateway die values as compared to the NGA, SBLB, SGA and NLDLB. It shows that the SCE-PSO algorithm does a proper load balancing operation, and it extends the lifetime of the network.

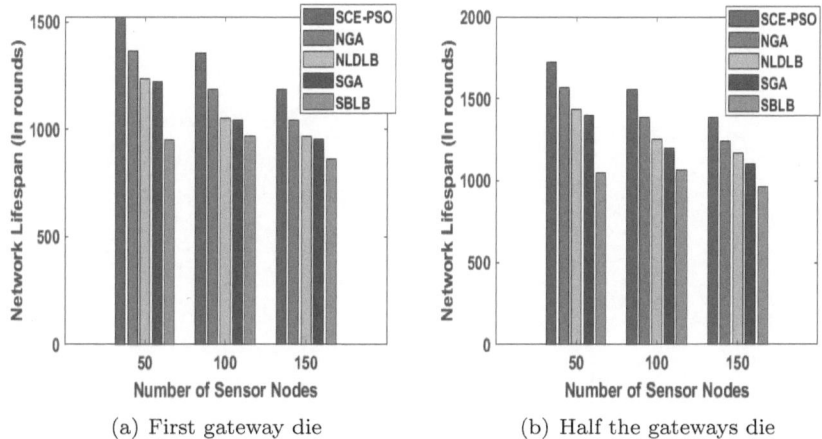

(a) First gateway die   (b) Half the gateways die

FIGURE 6.7: Comparison of network lifetime in terms of number of rounds using SGA, NGA, SBLB, NLDLB and proposed SCE-PSO algorithms for: (a) First gateway die; (b) Half the gateways die

6.6 Conclusion

In this chapter, a clustering approach based on SCE-PSO is proposed for efficient load balancing of gateways in WSNs. A novel fitness is also constructed by considering the massively loaded gateways and mean cluster distance. The SCE-PSO based load balancing technique has been compared with existing NGA, SBLB, SGA and NLDLB. The proposed SCE-PSO load-balancing algorithm performs well as compared with state-of-the-art NGA, SBLB, SGA and NLDLB algorithms. Experimental results show that the proposed solution has outperformed in terms of network lifetime, total energy utilization, the minimum lifetime gateway that dissolves its energy, and half the gateway die as compared to existing algorithms.

Chapter 7

PSO Based Routing with Novel Fitness Function for Improving Lifetime of WSN

7.1 Introduction

Over recent years, more attention has been paid to bio-inspired algorithms with the advancement of soft computing. Many researchers have developed algorithms to solve problems of WSNs, for example, the Genetic Algorithm(GA)[91], which is used to extend the lifetime of WSNs in large-scale surveillance applications. Artificial fish schooling algorithm [92], which has been used to solve the traffic monitoring system and fruit fly optimization algorithm [93], is applied in sensor node deployment. In this chapter, we propose a PSO based routing algorithm to address the energy efficiency in WSNs.

7.2 Preliminaries

7.2.1 Background of PSO

The significance of bird flocking inspires PSO [88], [94]. Birds fly together regularly in a group without colliding in search of shelter or food. All members or birds in a group follow the information of the group by changing its speed and position. Thus, the individual effort of the bird or member to find food and shelter is reduced in a group due to the sharing of group information. Let PSO consist of a specified number of particles (Pop_{size}) in the population, and each particle provides the solution to the specified problem case. Every particle in the population is calculated by a fitness function to determine the nature of the solution. The dimension of each particle in the population is equal. Particle P_i has a position (x_{id}) and velocity (v_{id}) in d^{th} dimension of the hyperspace. So the particle P_i is interpreted at any point of time as summarized as follows:

$$P_i = \{x_{i1}, x_{i2}, x_{i3}, \dots, x_{id}\}$$

In PSO, each particle P_i follows its own best (L_best_i) and the global (G_best) to adjust its location and velocity recursively in order to accomplish the global best position. L_best_i is the best position of particle P_i and G_best is the best solution among all the particles in the population. The below equations show the position and velocity updation of particle P_i in dimension d.

$$v_{id}(t+1) = w*v_{id}(t)+a_1*r_1*(L\_best_{id}-x_{id}(t))+a_2*r_2*(Gbest_d-x_{id}(t)) \quad (7.1)$$

$$x_{id}(t+1) = x_{id}(t) + v_{id}(t+1) \quad (7.2)$$

where w denotes the inertial weight, the constants of acceleration are a_1 and a_2, which are non-negative real numbers. r_1 and r_2 are two random numbers within the $[0,1]$ range that are distributed uniformly. The process of updating is repeated until we find the best global solution, or the maximum number of iterations obtained. The entire process of the PSO is shown in Figure 7.1.

7.2.2   Terminologies

The following terminology is used in the proposed algorithm and fitness function to explain the proposed work clearly.

1. The gateway set is $\lambda = \{g_1, g_2, \ldots, g_N\}$ and S_{sink} denotes the sink in the network.

2. $L(g_x)$ represents the lifetime of gateway g_x. If g_x has residual energy $ER(g_x)$ and energy usage per round $EU(g_x)$ then $L(g_x)$ is computed as follows:

$$L(g_x) = \left\lfloor \frac{ER(g_x)}{EU(g_x)} \right\rfloor$$

3. d_{max} shows maximum contact distance of each gateway.

4. $distance(g_x, g_k)$ shows the distance between the g_x and the g_k gateways.

5. $range(g_x)$ is the number of gateways within contact range of g_x (sometimes sink (S_{sink}) is also a contact range of $range(g_x)$). In other words, it is defined as follows:

$$range(g_x) = \{g_k | \forall g_k \in (\lambda + S_{sink}) \wedge distance(g_x, g_k) \leq d_{max}\} \quad (7.3)$$

6. $Relay\_Node$: $Relay\_Node$ is one of those network gateways. All of the network gateways do not reach the sink (S_{sink}) directly, in which case the gateways send data via other gateways, and this intermediate node is called $Relay\_Node$.

7. $Relay\_Nodes(g_x)$: It is a set containing a collection of gateways which can be allotted as a relay node for g_x for communicating with the sink S_{sink}.

$$Relay\_Nodes(g_x) = \{g_k | \forall g_k \in (range(g_x) - S_{sink})\} \quad (7.4)$$

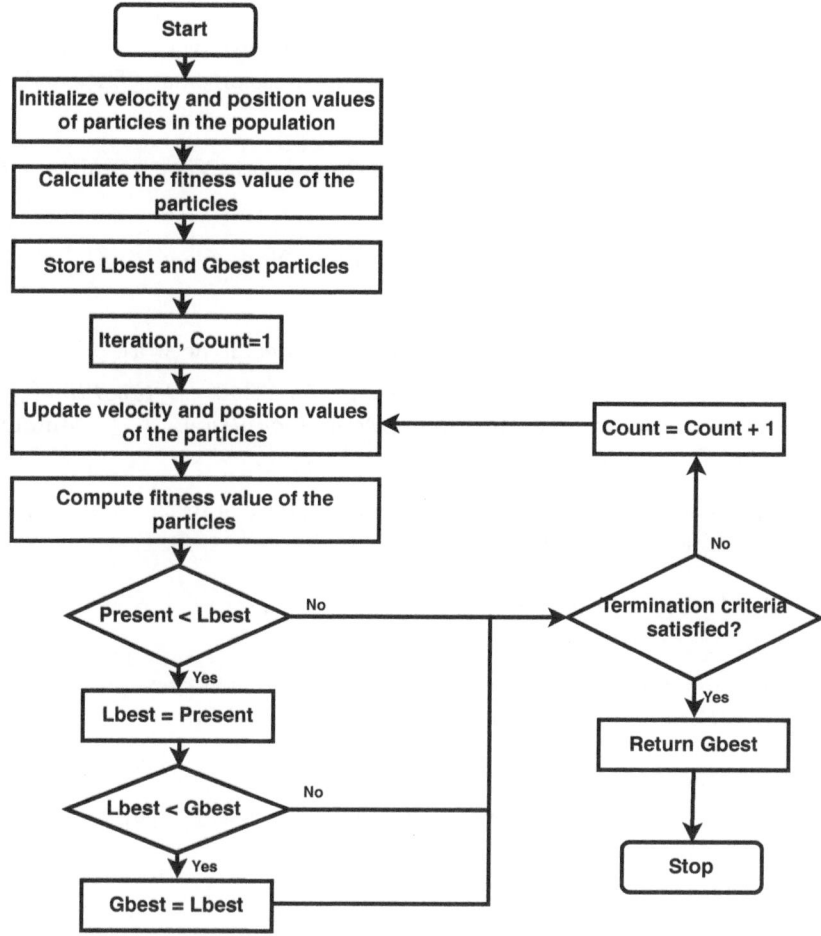

FIGURE 7.1: Flowchart of the PSO

8. $Relay\_Node(g_x)$: It denotes one of the gateways chosen as a relay node from g_x to S_{sink}. If g_i is within the contact range of S_{sink}, then we can communicate directly with S_{sink}.

9. $Relay\_Node\_Count(g_x)$: It indicates the number of relay nodes required to reach from g_x to the S_{sink}. If g_x is directly in contact with S_{sink} then $Relay\_Node\_Count(g_x)$ is one.

$$Relay\_Node\_Count(g_x) = \begin{cases} 1, & Relay\_Node(g_x) = S_{sink} \\ 1 + Relay\_Node\_Count(g_k), & \\ & Relay\_Node(g_x) = g_k \end{cases}$$
$$(7.5)$$

10. $Delay\_Time(g_x)$: It is time required to transfer the gathered data from the gateway (g_x) to the sink (S_{sink}). It includes average queuing delay (DT_q) for intermediate data delivery, transmission delay (DT_t) and propagation delay (DT_p).

$$Delay\_Time(g_x) = (DT_q + DT_t + DT_p) * Relay\_Node\_Count(g_x) \quad (7.6)$$

i.e.,

$$Delay\_Time(g_x) = C * Relay\_Node\_Count(g_x) \quad (7.7)$$

where
$C = DT_q + DT_t + DT_p$, is constant for a specific network [95].

It is also evident from Equation 7.7 that $Delay\_Time(g_x)$ is directly proportional to $Relay\_Node\_Count(g_x)$, i.e., Minimizing $Relay\_Node\_Count$ decreases delay time.

11. The total distance covered by any two gateways in the routing path is determined by $MaxDist$

$$MaxDist = \max\{distance(g_x, Relay\_Node(g_x)) | \forall x, 1 \le x \le N, g_x \in \lambda\} \quad (7.8)$$

12. The maximum relay node count of the gateway is defined as Max_Relay_Count. It has been formulated as follows:

$$Max\_Relay\_Count \\ = \max\{Relay\_Node\_Count(g_x) | \forall x, 1 \le x \le N, g_x \in \lambda\} \quad (7.9)$$

7.3 Proposed PSO based routing algorithm

In this study, we present a PSO based routing algorithm to enhance the lifetime of the WSNs. The PSO based routing algorithm consists mainly of three phases: the particle initialization phase, the fitness evaluation phase and the position and velocity update phase. The summary of the steps, as described above, is given below.

7.3.1 Random particle initialization phase

Each particle is encoded as a routing network. The size d of each particle P_i is equal to the number of gateways in the system. The addition or deletion of the gateway can lead to re-routing the network. The position of each gateway g_i is initialized with a randomly generated number from a uniform distribution in the range $(0, 1]$. The position value of dth gateway (i.e., x_{id}) assigns a new node (say g_k) to g_d as an immediate neighbour towards S_{sink}. That is, g_d sends data to g_k which sends data to S_{sink}. This can be formulated as follows:

$$n = ceiling(|x_{id} * Relay\_Nodes(g_d)|) \tag{7.10}$$

i.e., g_k is the nth relay node in $Relay\_Nodes(g_d)$. The random process of particle initialization is demonstrated by Example 1.

Example 1: Find a network of 12 gateways $\{g_1, g_2, g_3, \ldots, g_{12}\}$ as shown in Figure 7.2. The same is expressed by Table 7.1 in terms of gateways, relay nodes in the range and number of relay nodes in the region. Figure 7.2 shows the directed acyclic graph G(V, E). In G, V denotes the set of gateways and E represents set of edges. The edge between g_x and g_y specifies that g_x can send data to sink via g_y (g_y should be in communication range of g_x). It is clear from Figure 7.2 that g_2 can use any of the gateways from $\{g_4, g_5, g_6\}$ as next gateway to send data to S_{sink}.

During initialization, the position of each gateway (say x_{id}) of particle P_i is initialized with a random number in (0, 1] range. The allocated values for the random particle P_i are shown in Table 7.2. This random initialization of the gateway position values generates the solution to the problem. The generation of the random solution is as follows:

Let us consider the g_2 gateway and its given random value of 0.65. Based on the equation 7.10, $ceiling(0.65 * 3) = 2$, i.e., the second gateway (g_5) from the $Relay\_Nodes(g_2)$ is selected for data transmission to S_{sink}. This process is repeated for all gateways to create a whole network (particle). The final result of the network construction is shown in Table 7.2 and the same is shown in Figure 7.3.

7.3.2 Proposed fitness function

A novel fitness function is proposed to determine the goodness of the generated particles. The novel fitness function is given in the Equation 7.11. It is

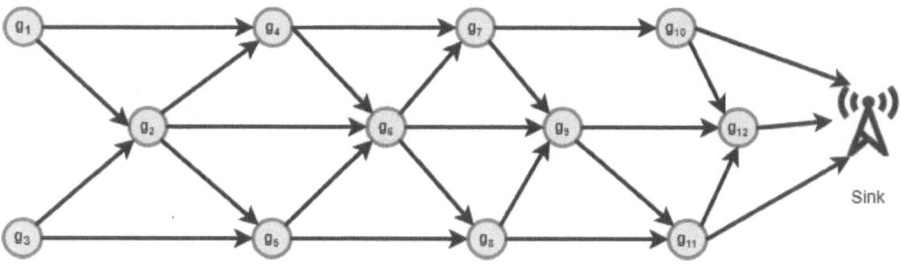

FIGURE 7.2: Random WSN scenario with 12 gateways and sink

TABLE 7.1: Gateways and their communication nodes

Gateways	$Relay\_Nodes(g_d)$	$|Relay\_Nodes(g_d)|$
g_1	$\{g_2, g_4\}$	2
g_2	$\{g_4, g_5, g_6\}$	3
g_3	$\{g_2, g_5\}$	2
g_4	$\{g_6, g_7\}$	2
g_5	$\{g_6, g_8\}$	2
g_6	$\{g_7, g_8, g_9\}$	3
g_7	$\{g_9, g_{11}\}$	2
g_8	$\{g_9, g_{10}\}$	2
g_9	$\{g_{10}, g_{11}, g_{12}\}$	3
g_{10}	$\{g_{12}, B_s\}$	2
g_{11}	$\{g_{12}, B_s\}$	2
g_{12}	$\{B_s\}$	1

TABLE 7.2: Relay node selection for data transfer from randomly generated particle

Gateways	x_{id}	$|Relay\_Nodes(g_d)|$	$ceiling(|x_{id} * Relay\_Nodes(g_d)|)$	$Relay\_Node(g_d)$
g_1	0.43	2	1	g_2
g_2	0.65	3	2	g_5
g_3	0.28	2	1	g_2
g_4	0.32	2	1	g_6
g_5	0.68	2	2	g_8
g_6	0.98	3	3	g_9
g_7	0.22	2	1	g_9
g_8	0.72	2	2	g_{10}
g_9	0.57	3	2	g_{11}
g_{10}	0.16	2	1	g_{12}
g_{11}	0.35	2	1	g_{12}
g_{12}	0.19	1	1	B_s

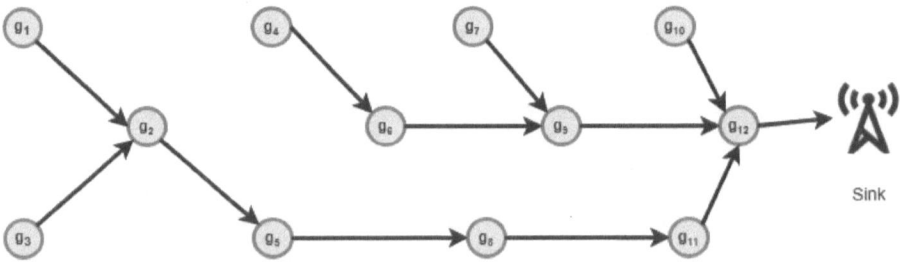

FIGURE 7.3: Routing solution for random particle

evident from the equation that the intended fitness function consists of considering the three goals. The first goal is to minimize the maximum distance between the gateway and the sink. The second one is to reduce the number of relay nodes used by the gateway for the transmission of data. The third objective is to minimize the relay load factor of the network.

$$Fitness = minimize \ \{\alpha * f_1 + \beta * f_2 + \gamma * f_3\} \qquad (7.11)$$

$$f_1 = minimize \ \{distance(g_i, Relay\_Nodes(g_i))\}, 1 \le i \le N \qquad (7.12)$$

$$f_2 = minimize \ \{Relay\_Node\_Count(g_i)\}, 1 \le i \le N \qquad (7.13)$$

$$f_3 = minimize \ \{Relay\_Load\_Factor\} \qquad (7.14)$$

$Relay\_Load\_Factor$ = sum of all relay nodes $Relay\_Load$ value whose $Relay\_Load$ value is above the Avg_Relay_Load value, i.e.,

$Relay\_Load\_Factor$

$$= \sum_{i=1}^{N} \{Relay\_Load(g_i) | Relay\_Load(g_i) \ge Avg\_Relay\_Load\} \qquad (7.15)$$

$Relay\_Load(g_i)$ = number of gateways sending data through g_i either directly or indirectly. $Relay\_Load(g_i)$ is zero, if gateway g_i is not participating as relay node.

$$Avg\_Relay\_Load = \left\lfloor \frac{\sum_{i=1}^{N} \{Relay\_Load(g_i)\}}{Number \ of \ relay \ nodes} \right\rfloor \qquad (7.16)$$

where α, β and γ are weighted importance of each objective such that $\alpha + \beta + \gamma = 1$ and $0 \le \alpha, \ \beta, \ \gamma \le 1$.

For a better solution, the fitness function needs to be minimized, i.e., lower fitness value indicates the best network.

The fitness value calculation is explained using Example 2.

Example 2:
Let us consider from Figure 7.3 the relay nodes are $\{g_2, \ g_5, \ g_6, \ g_8, \ g_9, \ g_{11}, \ g_{12}\}$.
The $Relay\_Load$ values of $\{g_2, \ g_5, \ g_6, \ g_8, \ g_9, \ g_{11}, \ g_{12}\}$ are $\{$ 2, 3, 1, 4, 3, 9, 11$\}$, respectively.
Then $Avg\_Relay\_Load = \lfloor 33/7 \rfloor = 4$ and $Relay\_Load\_Factor = 4 + 9 + 11 = 24$.

Remember that while achieving any of the above objectives, we can lose control of others. We should include all three objectives in a single fitness function and minimize them. To overcome the problem of trade-offs between objectives, we used a weighted sum approach to formulate the proposed fitness function by considering all goals.

7.3.3 Position and velocity updating phase

Every iteration updates the position and velocity of each particle using the Equation 7.1 and Equation 7.2. When updating the velocity and location of the particles, we can find a new position due to algebraic addition and subtraction. New position values can be less than or equal to zero, or greater than one. However, according to the Equation 7.10, the position of the particles must be within the range of (0, 1]. To get the right range values, we need to make the following improvements to our algorithm:

1. If the modified position value is less than or equal to zero, then change the value with a newly generated random number whose value tends to be zero.

2. If modified position value is above one, then change the value to one.

After obtaining the modified positions, the particle P_i is evaluated using a fitness function. Each particle best fitness value ($Lbest_i$) is amended by itself, only if its present fitness value is better than $Lbest_i$ fitness value. The velocity and position values are modified recursively until the termination condition is satisfied. After termination of the PSO based routing algorithm, the endmost solution is represented by the *Gbest*.

7.4 Results and discussion

The proposed approach has been implemented using the MATLAB 2015R and C++ language. All the simulations are conducted under two different WSN conditions, namely *wsn*1, *wsn*2. The *wsn*1 and *wsn*2 consist of $60 - 80$ gateways and $100 - 400$ sensor nodes. The deployed scenarios vary only in the topology of the network and the location of the sink. The locations of the sink are $(200, 200)$ m^2 and $(350, 150)$ m^2 respectively. Table 7.3 displays those standard simulation parameters for both scenarios *wsn*1 and *wsn*2. For comparison purposes GA [63], Greedy Load Balancing Clustering Algorithm (GLBCA) [96] and PSO [90] existing algorithms have been introduced in the literature. To validate the performance of the proposed method, it is compared with GA, GLBCA, PSO based routing algorithms in terms of network lifetime, the total number of hops and average relay load on the network.

7.4.1 Network lifetime vs number of gateways

Experiments are carried out on the number of sensor nodes $100 - 400$. The network lifetime of both the scenarios *wsn*1 and *wsn*2 is measured in terms of the number of rounds. The network lifetime is defined as the number of

---

Algorithm 4 PSO based routing

- **Input**

  - Gateway set $\lambda = \{g_1, g_2, ..., g_N\}$
  - S_n=Swarm size
  - Termination criteria

- **Output**

  - Optimal route from each gateway to the sink in the network
  - Routing R: $\lambda \rightarrow \{\lambda + S_{sink}\}$

- **Algorithm**

  1: **Step 1:**
  2: Initialize each particle (P_i) in the swarm (S_n), $\forall i, 1 \leq i \leq S_n$
  3: **Step 2:**
  4: **for** $i = 1$ to S_n **do**
  5:     Evaluate fitness value of each particle P_i, i.e., $Fitness(p_i) = \alpha * f_1 + \beta * f_2 + \gamma * f_3$
  6:     $L\_best_i \leftarrow P_i$
  7: **end for**
  8: **Step 3:**
  9: $G\_best = \{L\_best_x | Fitness(L\_best_x) = minimum(Fitness(L\_best_i),$ $\forall i, 1 \leq i \leq S_n)\}$
  10: **Step 4:**
  11: **while** (!$terminate$) **do**
  12:     **for** $i = 1$ to S_n **do**
  13:         Update position and velocity of particle P_i using Equation 7.2
  14:         Evaluate fitness value of P_i
  15:         **if** F$(P_i) <$ F(L_best_i) **then**
  16:             $L\_best_i \leftarrow P_i$
  17:         **end if**
  18:         **if** F$(L\_best_i) <$ F(G_best) **then**
  19:             $G\_best \leftarrow L\_best_i$
  20:         **end if**
  21:     **end for**
  22: **end while**
  23: **Step 5:**
  24: Store $Relay\_Node(g_i)$, $\forall i, 1 \leq i \leq N$ using G_best, i.e., Efficient routing path for the given network.
  25: **Step 6:**
  26: Stop

TABLE 7.3: Experimental parameters and their values

Parameters	Value
WSN parameters	
Target area	$500 * 500m^2$
Sensor nodes	$200 - 500$
Energy of sensor node	2.0J
Gateways	$60 - 80$
Energy of gateway	12.0J
Communication range	$150m$
E_{elec}	$50nJ/bit$
ϵ_{fs}	$10pJ/bit/m^2$
ϵ_{mp}	$0.0013pJ/bit/m^4$
E_{DA}	$5nj/bit$
Packet size	$4000bits$
Message size	$200bits$
Parameters of PSO	
S_n	50
Number of iterations	500
a_1	1.4962
a_2	1.4962
W	0.7968
V_{max}	0.5
V_{min}	-0.5

rounds the complete network is alive. The experimental results for $wsn1$ are shown in Figure 7.4, Figure 7.5 and Figure 7.6 for 60, 70 and 80 gateways. Similarly, $wsn2$ results for 60, 70 and 80 number of gateways are shown in Figure 7.7, Figure 7.8 and Figure 7.9, and the results show that the network lifetime of $wsn1$ and $wsn2$ using the proposed approach is extended relative to state-of-the-art PSO and GLBCA approaches.

7.4.2 Number of hops vs number of gateways

Experiments are carried out on the number of gateways 60, 70 and 80. For both $wsn1$ and $wsn2$, the number of hops is determined in terms of communication between the gateways. The experimental results for $wsn1$ are shown in Figure 7.10 for 60, 70 and 80 gateways. Similarly, the $wsn2$ results for 60, 70 and 80 gateways are shown in Figure 7.11. From the results, the proposed method often needs more routing hops compared to state-of-the-art algorithms such as GA and PSO.

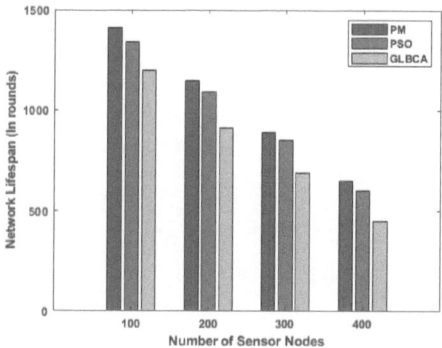

FIGURE 7.4: Comparison of network lifetime with 60 gateways and 100−400 sensor nodes for *wsn*1

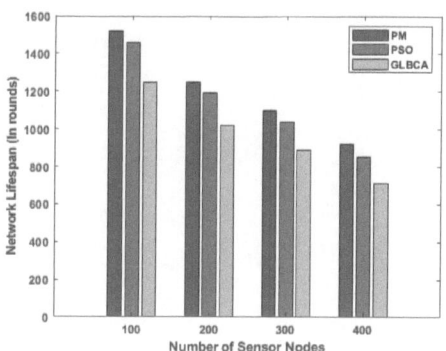

FIGURE 7.5: Comparison of network lifetime with 70 gateways and 100−400 sensor nodes for *wsn*1

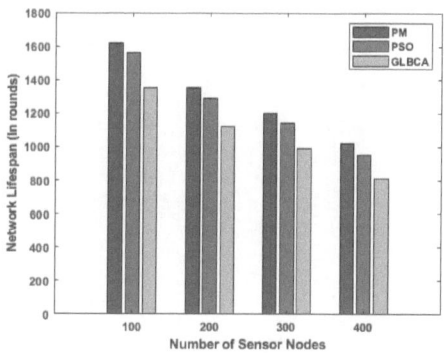

FIGURE 7.6: Comparison of network lifetime with 80 gateways and 100−400 sensor nodes for *wsn*1

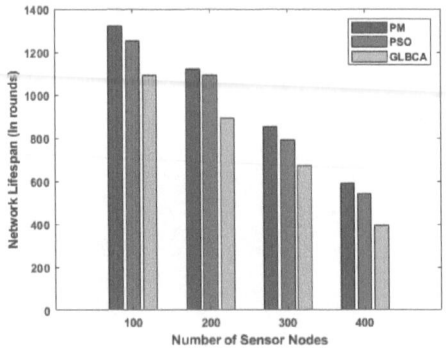

FIGURE 7.7: Comparison of network lifetime with 60 gateways and 100−400 sensor nodes for *wsn*2

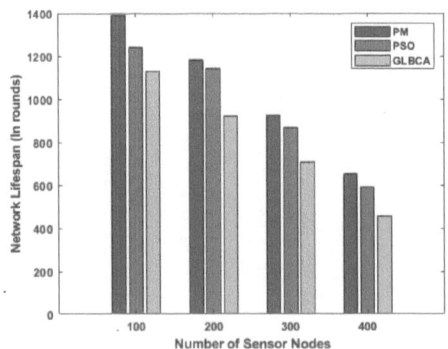

FIGURE 7.8: Comparison of network lifetime with 70 gateways and 100−400 sensor nodes for *wsn*2

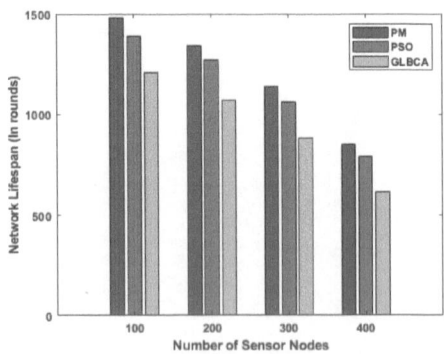

FIGURE 7.9: Comparison of network lifetime with 80 gateways and 100−400 sensor nodes for *wsn*2

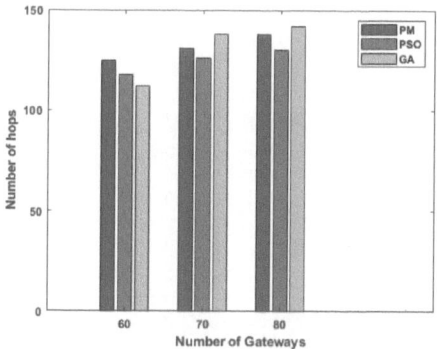

FIGURE 7.10: Observation of number of gateways vs number of hops in *wsn*1 for 60 − 80 gateways

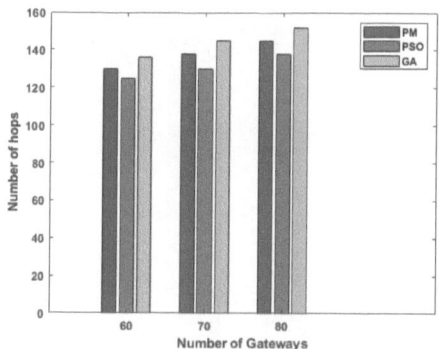

FIGURE 7.11: Observation of number of gateways vs number of hops in *wsn*2 for 60 − 80 gateways

7.4.3   Average relay load vs number of gateways

Experiments are conducted for the number of gateways 60, 70 and 80. For both *wsn*1 and *wsn*2, average relay load on the network is calculated as the mean of all relay loads in the network. These experimental results for *wsn*1 are shown in Figure 7.12 for 60, 70 and 80 number of gateways. Similarly, *wsn*2 results for 60, 70 and 80 number of gateways are shown in Figure 7.13. It is observed from the results the proposed method requires less average relay load when compared to state-of-the-art GA and PSO techniques. It is due to considering the relay load factor in the proposed fitness function.

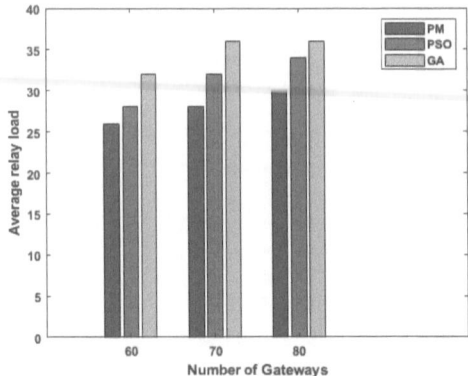

FIGURE 7.12: Comparative analysis in terms of number of gateways vs average relay load in $wsn1$ with $60 - 80$ gateways

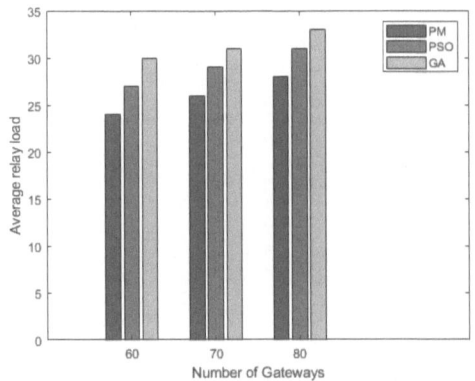

FIGURE 7.13: Comparative analysis in terms of number of gateways vs average relay load $wsn2$ with $60 - 80$ gateways

7.5 Conclusion

In this study, the PSO based routing algorithm is used for efficient routing in WSNs to prolong the lifetime of WSNs. For a suitable path, a novel fitness is also designed to take into account the distance, hop-count and relay load factor of the network. The proposed PSO-based routing algorithm is compared to state-of-the-art routing approaches such as GA and PSO and also GLBCA is used to analyze the network lifetime of the WSNs. The efficiency of the proposed algorithm is evaluated in terms of network lifetime, number of hops and average relay load under two separate WSN typologies. The experimental results show that the proposed PSO based routing algorithm increased the network lifetime and reduced the average relay load at the cost of slightly more number of hops when compared with state-of-the-art routing techniques.

Chapter 8

M-Curves Path Planning for Mobile Anchor Node and Localization of Sensor Nodes Using DSA

8.1 Introduction

Major applications of WSN including military surveillance, environmental monitoring, forest fire detection, health care monitoring, etc., need accurate location of the sensor node. For example, in military surveillance, sensor's precise location is required for detecting when the person from opponent team enters into surveillance area [28]. Localization can be termed as computing the position of the sensor node deployed in the network. Global Positioning System (GPS) can be used for localizing the sensor node. Due to limited power resource of sensor nodes, high cost and poor performance of GPS in the indoor environment, GPS is not an efficient solution for localization [29, 30].

Localization techniques can be broadly classified into range based and range free mechanisms. Range based techniques use distance/angle information to localize a node whereas range free techniques use connectivity information for localization. Some common range based techniques are Time of Arrival (ToA), Time Difference of Arrival (TDoA), Angle of Arrival (AoA) and Received Signal Strength Indicator (RSSI) [31]. Range free techniques are less complex, cost efficient and do not need any additional hardware but range based techniques give improved results with respect to localization accuracy [32].

In localization process, the sensor nodes which know their position in the deployed field are called *anchor nodes* or *reference nodes* or *beacon nodes*. Unknown nodes are the nodes which do not know their location in the network and are localized with the help of anchor nodes. Anchor nodes broadcast beacon signals periodically with their respective spatial coordinates to the unknown nodes for their localization; hence they consume more energy than the unknown sensor nodes. All the unknown nodes in the network rely on the information sent by the anchor nodes with which they can localize themselves with some localization techniques. The anchor nodes can be either static or mobile and are deployed in the network manually. The mobile anchor nodes are additionally equipped with GPS.

Static anchors are used in most of the localization approaches. The major drawback of this approach is, when the network is scaled to large network, the number of anchors needed for the network is high, which is not a cost effective solution as the energy consumption increases with increase in anchor nodes. Further, the anchor nodes do not have any usage in the network. Conversely, for some applications such as military missions, human operation is absurd. Considering all the issues mentioned, we are motivated to localize the sensor nodes with a single mobile anchor node in place of multiple static anchor nodes. Also, using mobile anchor for localization is a cost efficient solution and it outperforms in terms of accuracy than static sensor nodes [33, 34, 35, 36, 37]. A basic issue with the mobile beacon is finding the optimal trajectory with which mobile beacon node travels and sends beacon signal with its location information to other sensor nodes.

Hence, after this, mobile anchor based localization is bounded to estimating an optimum path for mobile anchor. To solve this problem, several properties should be considered for finding optimum trajectory of mobile anchor node. A necessary condition for optimum trajectory is that all unknown sensor nodes should receive sufficient beacon signals (at least three for 2-dimensional network) to estimate their position. The important properties that a trajectory should hold are discussed in [33, 34, 35]. The basic property is having shortest path achieving full coverage with messages from non-collinear anchors. Various trajectories for mobile anchor have been proposed in [38, 39, 40, 41]. Each of these trajectories is different from one another in the pattern which the trajectory follows. Localization accuracy and path length of the trajectory are improved in most of these papers. The works related to the model are discussed in the succeeding paragraphs.

8.2   Preliminaries

8.2.1   Overview of dolphin swarm algorithm

The dolphin is one of the smartest animals which has significant and interesting biological characteristics. Also they use their smartness in living habits like communicating, hunting the prey, etc. Its intelligence and biological characteristics are associated with swarm intelligence and it is used to solve optimization problems. This swarm optimization algorithm is proposed as Dolphin Swarm Algorithm (DSA) in [50]. Some interesting intelligent characteristics of dolphins used for DSA are echolocation, cooperation and division of labor and information exchange which are discussed below.

Echolocation: Though a dolphin has good eyesight, it does not help significantly during predation in poor light conditions. Hence a dolphin uses its special intelligence called echolocation to search for prey. With echolocation, it makes sounds to find the prey and with the intensity of the echo it can estimate the location, distance and size of prey. So echolocation helps dolphins have a better knowledge about their surrounding environment.

Cooperation and division of labor: Predation is mostly done by the joint efforts of the dolphins with cooperation and division of labor. Attacking a larger prey cannot be done by a single dolphin. Hence the dolphin calls other dolphins for help and with cooperative behaviour of all the dolphins they attack the prey.

Information exchanges: Dolphins use different frequencies of sound to exchange the information within their own language system. In cooperation and division of labor phase, they use this special ability to call the other dolphins and update the location of the prey. This information exchange between dolphins enhances the actions which are taken by dolphins during the predation phase.

8.2.2 Terminologies

Terminologies of DSA used in [50] are discussed in the following paragraphs.

a) **Dolphin:**
   Using swarm intelligence to solve the optimization problem, a set of dolphins is considered to be a set of feasible solutions. Dolphin's predatory behaviour is used to solve the problem of optimization. Each feasible solution in a D dimensional field is defined as a dolphin, $Dolph_i$ $=[x_1, x_2, x_3, \ldots, x_D]^T$ (i=1,2,...,N), where N is the number of dolphins and x_j is the component of each dimension to be optimized.

b) **Individual optimal solution and neighbourhood optimal solution:**
   Individual optimal solution and neighbourhood optimal solution are the two variables represented by $Ibest_i$ (i=1,2,...,N) and $Nbest_i$(i=1,2,...,N), respectively, which are associated with each dolphin $Doplh_i$ (i=1,2,...,N). $Ibest_i$ has the optimal solution of each dolphin in each iteration and $Nbest_i$ represents the optimal solution of each dolphin selected from its neighbourhood and itself.

c) **Fitness**
   Fitness function is used to evaluate whether the given solution is better or not. In DSA, when the value of fitness is close to zero, the solution is considered to be a better solution.

d) **Distances**
   In DSA, three types of distance measures are defined. The distance between two dolphins $Dolph_i$ and $Dolph_j$ is termed as $DDol_i, j$, and is given by

$$DDol_{i,j} = ||Dolph_i - Dolph_j||, i, j = 1, 2, \ldots, n, i \neq j \qquad (8.1)$$

The distance between the optimal solution $Nbest_i$ and the dolphin $Dolph_i$ is termed as $DNbest_i$, and is given by

$$DNbest_i = ||Dolph_i - Nbest_i||, i = 1, 2, \ldots, n \qquad (8.2)$$

The distance between the optimal solution $Nbest_i$ and the dolphin $Ibest_i$ is termed as $DINbest_i$, and is given by

$$DINbesti = ||Ibesti - Nbesti||, i = 1, 2, \ldots, n \qquad (8.3)$$

8.2.3   Phases of DSA

DSA is divided into initialization, search, call, reception, predation and termination phases[50]. In this subsection, search phase, call phase, reception phase and predation phase are described.

(i) **Search phase** In the search phase, each dolphin searches its surrounding area by producing sounds in K random directions. The sound produced by a dolphin is termed as $V_i = [v_1, v_2, v_3, \ldots, v_K]^T$ ($i = 1, 2, 3, \ldots, n$),where K is the number of sounds and v_j represents the attributes for direction of the sound. $||V_i|| = $ sp (i=1,2,3,...,n), where sp is the constant representing speed of the sound. T is the maximum search time for the dolphins. Each dolphin $Dolph_i$ searches a new solution Sol_{ijt}, by making sounds V_i at time t, which is given by

$$Sol_{ijt} = Dolph_i + V_{jt} \qquad (8.4)$$

Fitness for the new solution Sol_{ijt} of each dolphin $Dolph_i$ is calculated by

$$F_{ijt} = Fitness(Sol_{ijt}) \qquad (8.5)$$

The solution with minimum fitness value is calculated and the corresponding solution is assigned to the individual optimal solution $Ibest_i$.

$$F_{iab} = min_{j=1,2,3,\ldots,K;t=1,2,3,\ldots,T}\{F_{ijt}\} \qquad (8.6)$$

$$Ibest_i = Sol_{iab} \qquad (8.7)$$

Individual optimal solution $Ibest_i$ is assigned to neighbourhood optimal solution $Nbest_i$,

if
$$Fitness(Ibest_i) < Fitness(Nbest_i) \qquad (8.8)$$

else $Nbest_i$ does not change.

(ii) **Call phase** In the call phase, each dolphin makes sounds to inform about the solution found in the search phase. Also it informs about the best solution obtained in the search phase and its location. A matrix of size $N \times N$ called 'Transmission time matrix'(TS) is used to control the information exchange process, where $TS_{i,j}$ represents the time elapsed for the sound to reach from $Dolph_i$ to $Dolph_j$. For the element $TS_{i,j}$ from the transmission time matrix and for any two neighbourhood solutions $Nbest_i$ and $Nbest_j$

if

$$Fitness(Nbest_i) < Fitness(Nbest_j) \tag{8.9}$$

and

$$TS_{i,j} > \lceil \frac{DDol_{i,j}}{A.sp} \rceil \tag{8.10}$$

then $TS_{i,j}$ is updated as

$$TS_{i,j} = \lceil \frac{DDol_{i,j}}{A.sp} \rceil \tag{8.11}$$

where A is a constant which is the acceleration of sound, controls the speed of the sound and sp is the speed as aforementioned. Once updation of the matrix TS is done, the algorithm enters into the reception phase.

(iii) **Reception phase** In the reception phase, as an indication of sound being spread in one unit of time, all the elements of $TS_{i,j}$ are decreased by one. The algorithm checks the elements of the matrix TS and the sound sent by $Dolph_i$ can be received by $Dolph_j$, if

$$TS_{i,j} = 0 \tag{8.12}$$

Then to demonstrate the reception of sound, the particular element $TS_{i,j}$ is replaced by the term named 'Maximum search time' (T_2). The neighbourhood solution is updated if it has received any best solution by comparison using

$$Fitness(Nbest_i) < Fitness(Nbest_j) \tag{8.13}$$

(iv) **Predation phase** Each Dolphin has a set of information which has its own position $Dolph_i$ in the D-dimensional space, its individual optimal solution $Ibest_i$, its neighbourhood optimal solution $Nbest_i$, the distances DDol, DNbest, DINbest. The search radius R_1, which is the maximum range a dolphin can search for the solution, can be calculated by

$$R_1 = T_1 \times sp \tag{8.14}$$

In the predation phase, each dolphin calculates the radius R_2 which is the distance between dolphin's neighbourhood optimal solution and the optimal position after the predation phase using the set of known information. The encircling radius R_2 is calculated based on three cases.

Case 1. $DNbest_i \leq R_1$

In this case, neighbourhood optimal solution is within the range of the $Dolph_i$; then new range R_2 is calculated using

$$R_2 = \left(1 - \frac{2}{e}\right) * DNbest_i \tag{8.15}$$

where e is the 'radius reduction coefficient' which is a constant used to control the radius R_2 and makes it converge to zero. Once the radius is calculated, new position for dolphin $Dolph_i$ is calculated using

$$newDolph_i = K_i + \frac{Dol_i - Nbest_i}{DNbest_i} R_2 \tag{8.16}$$

Case 2. $DNbest_i > R_1$ and $DNbest_i \geq DINbest_i$

In this case, $Ibest_i$ is closer to $Dolph_i$ and its $Nbest_i$ is updated with the information received from others; then new range R_2 is calculated using

$$coeff_N = \frac{\frac{DNbest_i}{Fitness(Nbest_i)} + \frac{DNbest_i - DINbest_i}{Fitness(Ibest_i)}}{e.DNbest_i \frac{1}{fitnessNbest_i}} \tag{8.17}$$

$$R_2 = \left(1 - coeff_N\right) * DNbest_i \tag{8.18}$$

Once the radius is calculated, new position for dolphin $Dolph_i$ can be calculated by moving it to a random position which is R_2 distance away from $Nbest_i$ which can be calculated using

$$newDolph_i = Nbest_i + \frac{Random}{||Random||} R_2 \tag{8.19}$$

Case 3. $DNbest_i > R_1$ and $DNbest_i < DINbest_i$

In this case, $Nbest_i$ is closer to $Dolph_i$ and its $Nbest_i$ is updated with the information received from others; then new range R_2 is calculated using

$$coeff_{IN} = \frac{\frac{DNbest_i}{Fitness(Nbest_i)} + \frac{DINbest_i - DNbest_i}{Fitness(Ibest_i)}}{e.DNbest_i \frac{1}{fitnessNbest_i}} \tag{8.20}$$

$$R_2 = \left(1 - coeff_{IN}\right) * DNbest_i \tag{8.21}$$

Once the radius is calculated, new position for dolphin $Dolph_i$ is evaluated by moving it to a random position which is R_2 distance away from $Nbest_i$ which can be calculated using Eq. 8.19.

After the $Dolph_i$ changes its position to $newDolph_i$, the fitness is calculated and compared with the neighbourhood optimal solution $Nbest_i$

$$Fitness(newDolph_i) < Fitness(Nbest_i) \qquad (8.22)$$

If the above condition is satisfied, the algorithm replaces $Nbest_i$ by $newDolph_i$ else $Nbest_i$ does not change. The phases of DSA are explained graphically with in Figure 8.1.

8.2.4   DSA for localization

Although various optimization algorithms such as Particle Swarm Optimization (PSO)[51], Artificial Bee Colony algorithm (ACO)[52], Bacterial Foraging algorithm(BFA)[53] were proposed in the literature for the problem of localization, DSA is chosen for optimization in this work because of its interesting features. DSA gives better results in lower dimension unimodal functions and higher dimension multimodal functions. As the proposed work is in two dimensional model, DSA is opted for optimization. Also DSA possess first-slow-then-fast and periodic convergence. The dolphin searches for more time in its own surrounding area, which extends the search area of dolphin swarm at group level [50]. This feature avoids premature convergence in the optimization problem. Hence, for optimizing the fitness function used for localization, DSA is adopted.

8.2.5   System models

The proposed path planning model is designed with the following considerations.

1. A wireless sensor network of two dimensional field is deployed in a square region with area L m^2.

2. A set of unknown sensor nodes are deployed randomly in the field with uniform distribution. These unknown nodes are assumed to be static throughout the process.

3. Initially, the unknown node does not know its location in the deployed field.

4. Each sensor node in the network has a fixed transmission range tr m.

5. The mobile anchor node moves freely inside the network in straight lines depending on the path model. The mobile anchor nodes are location aware inside the network.

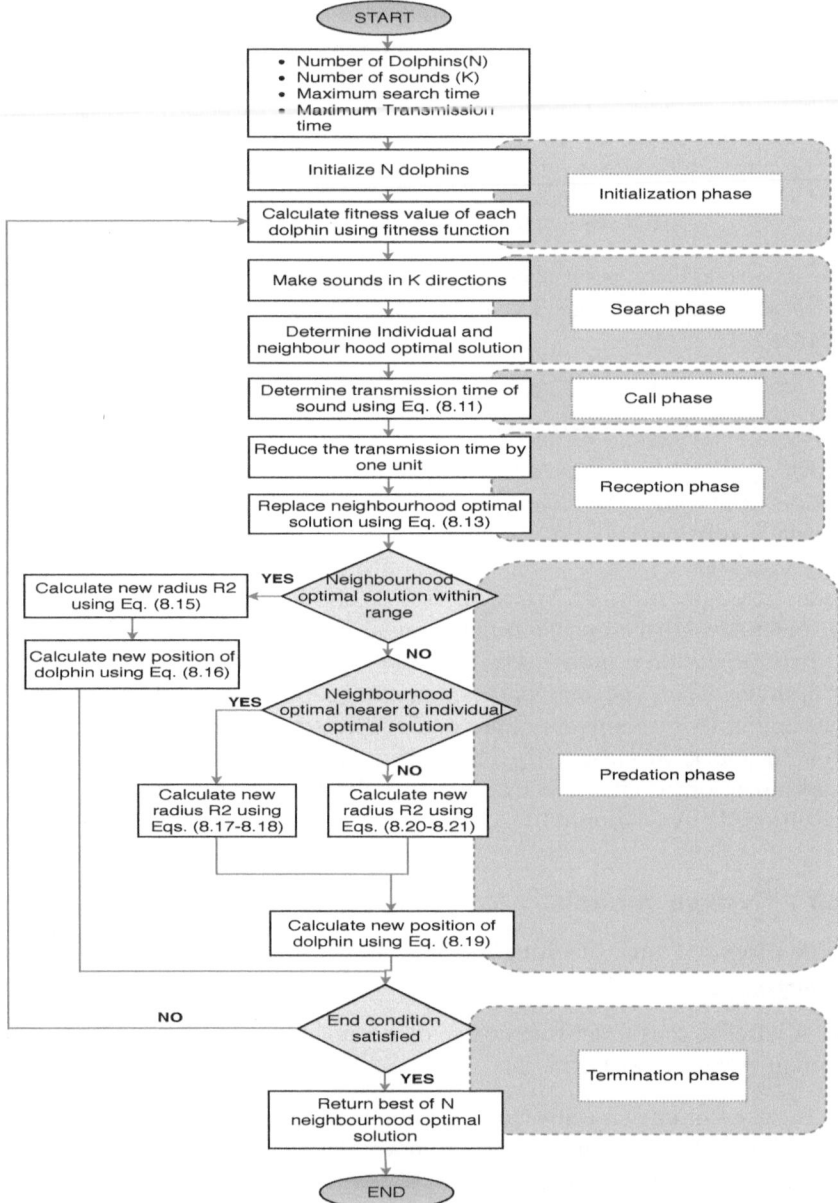

FIGURE 8.1: Flowchart of the phases of Dolphin Swarm Algorithm

6. During the anchor movement phase, it stops at particular points in its path which are termed *anchor points*. Anchor node transmits beacon signals to the other nodes within its transmission range from the anchor points.

7. The distance between two consecutive anchor points in the trajectory is set as dt m.

8. Each mobile node and anchor node communicate within themselves iff both the nodes lie in their transmission ranges.

9. An unknown node can start the process of localization using any localization algorithm if it receives three different beacon messages.

10. The mobile anchor node consumes more energy than the unknown node as it traverses in the network and transmits a beacon signal.

8.2.6   Localization technique

Trilateration is the most common technique used for estimating the position of unknown nodes in WSN. Each unknown node receives beacon messages from the anchor node [54]. Once an unknown node receives sufficient number of beacon messages, it will start the process of localization. Each unknown node will calculate its respective distances with the received beacon messages and position of node is calculated using trilateration. Consider an unknown sensor node deployed in 2D environment at position (x_i, y_i) and the anchor nodes positioned at $(X_1, Y_1), (X_2, Y_2), (X_3, Y_3)$ where $dist_1$, $dist_2$, $dist_3$ are the respective distances from the anchor nodes and the unknown sensor node, then by solving the following equations, the position of S_i can be estimated.

$$dist_1 = \sqrt{((X_1 - x_i)^2 + (Y_1 - y_i)^2)}$$
$$dist_2 = \sqrt{((X_2 - x_i)^2 + (Y_2 - y_i)^2)} \qquad (8.23)$$
$$dist_3 = \sqrt{((X_3 - x_i)^2 + (Y_3 - y_i)^2)}$$

Accuracy-Priority Trilateration (APT) is a technique which estimates the location of sensors by considering three nearest beacon messages among the received messages. This technique provides high localization accuracy compared to other trilateration techniques [39]. APT is considered for the evaluation of the performance of the Localization process using DSA.

8.3   Proposed work

8.3.1   Problem formulation

Given a region R, where the wireless sensor network is deployed with a set of static sensor nodes $SS = \{SS_1, SS_2, SS_3..., SS_n\}$ and a mobile anchor node MA, the problem is to find an optimal trajectory for the mobile anchor node. The optimality of the trajectory is termed by the properties such as full

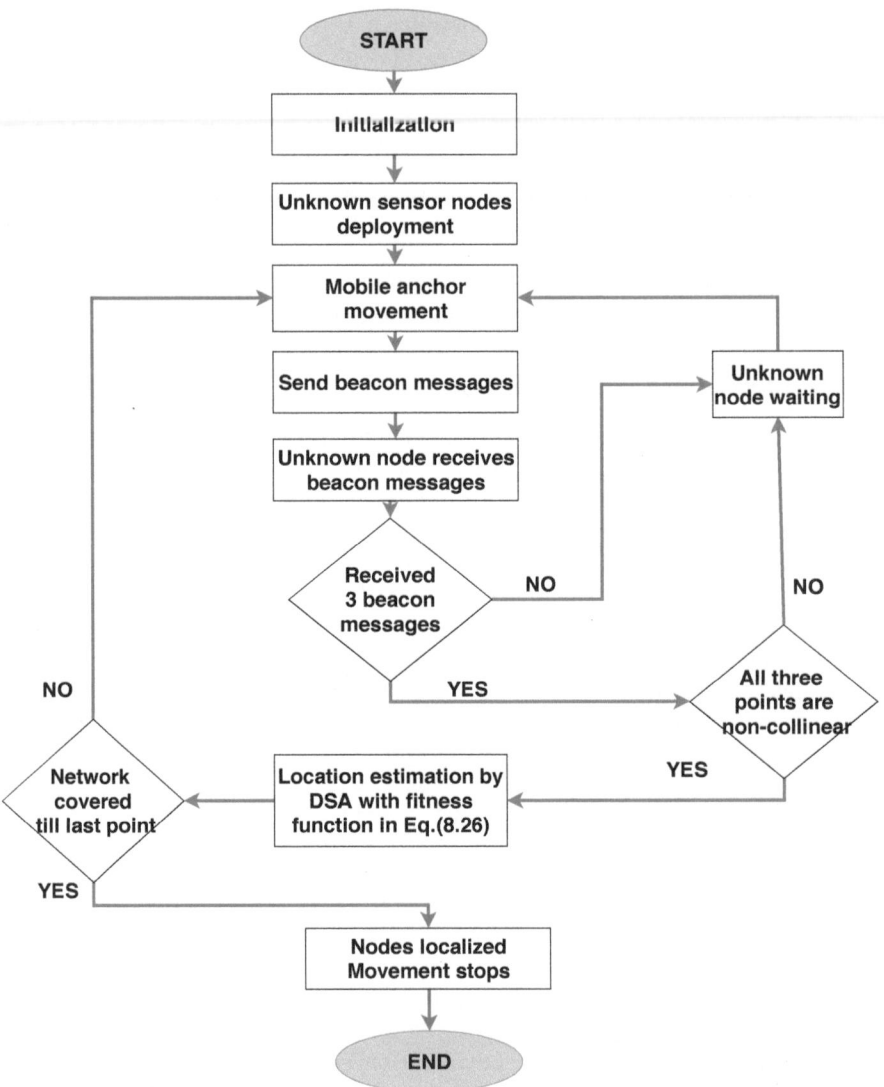

FIGURE 8.2: Flowchart of node localization process

coverage, shortest path length and high localization accuracy. Consider the mobile anchor node. MA is at location (X, Y) known as anchor point and a static sensor node, SS_i, is at position (Sx_i, Sy_i); then node MA can be utilized for the localization process of SS_i iff both the nodes are in the transmission range tr. That is,

$$\sqrt{((X - SSx_i)^2 + (Y - SSy_i)^2)} \leq tr \qquad (8.24)$$

8.3.2 Mobile anchor movement

With the aforementioned properties of the optimal trajectory of mobile anchor node, a path planning mechanism is designed and called "M-Curves" as the trajectory follows the pattern 'M' as shown in the Figure 8.3. The trajectory assures that each static sensor node receives at least three non-linear beacon messages for the efficient localization.

In spite of several path planning models available in the literature, each model has its own advantages and disadvantages. Path length, collinearity of the mobile anchor points and coverage area are the major features to be considered. Path length was considered for the design of two models, SCAN and HILBERT. The localization accuracy of the models is less compared to other methods because of the collinearity and coverage problems. The collinearity and coverage issue were concerns for the LMAT model. However, the path length of the model is too high in comparison with any other models. All the three major features are considered for design in Z-Curves. The anchor node needs higher transmission range for covering the edges of the network in Z-Curves model. Also, the localization accuracy can be further improved. The model Z-curves satisfies all features and attains better localization accuracy compared to other models.

Mobile anchor path is designed by considering the basics of trilateration. Figure 8.4 shows the basic M-Curves of the proposed trajectory. The total length taken by the basic M-Curves is $2dt + \sqrt{2}dt$. The network is divided into four sub-squares, namely $square_{sq}$, sq=(1,...,4) and the center of each sub-square is given by C_{sq}. C_0 represents the center of the network which is also the center of the basic curve. Each unknown sensor node should receive

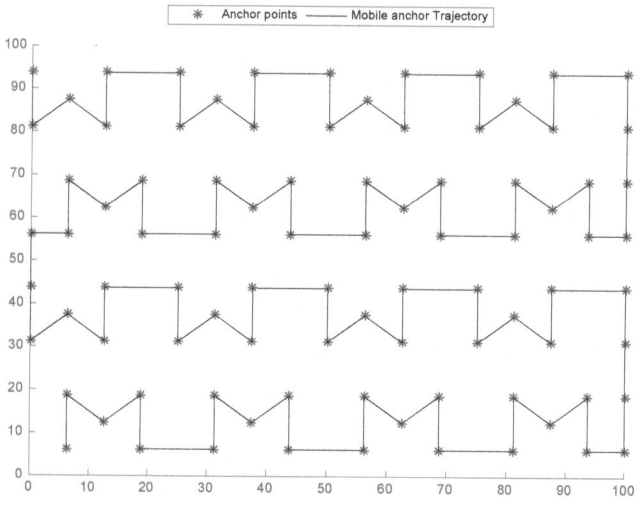

FIGURE 8.3: Proposed mobile anchor trajectory: M-Curves

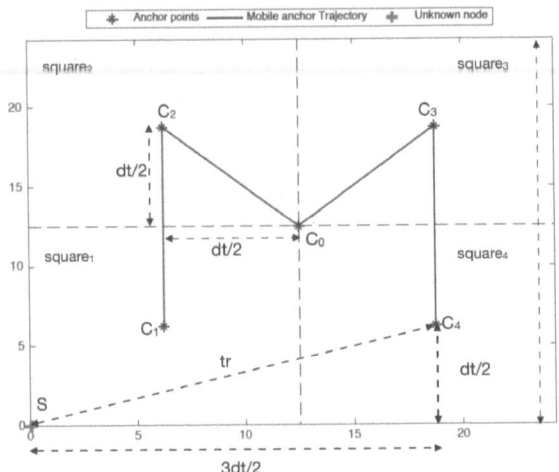

FIGURE 8.4: Basic M-Curve

three beacon messages for successful localization. Hence the beacon messages transmitted at C_{sq} should be received by the nodes in the $square_{sq}$ and in addition to that, nodes in the two adjacent sub-squares should receive the beacon messages such that each sensor node in the network receives at least three beacon messages for successful localization. With the conditions given, the transmission range of the anchor node tr can be found by considering the scenario given in the Figure 8.4. Consider an unknown sensor node S which is the farthest sensor node from C_4. If the node S receives beacon message from position C_4, then the beacon message sent from C_4 is assured to be received by all the unknown nodes inside $square_1$, $square_3$ and $square_4$. Therefore transmission range is set to the distance between the center of the sub-square to the farthest node in the adjacent sub-squares.

$$tr = \sqrt{\left(\frac{dt}{2}\right)^2 + \left(\frac{3dt}{2}\right)^2}$$

$$tr = \sqrt{\frac{5}{2}} dt$$

Hence, to ensure that all the unknown sensor nodes in the network receive atleast three beacon messages, transmission range is set such that $tr \geq \sqrt{\frac{5}{2}} dt$.

Starting from its initial location, the anchor node traverses a path which seeks the pattern of the alphabet 'M'. The pattern 'M' is depicted in such a way that each of the vertical side segments forming the alphabet has equal length dt and the diagonal line segments of the 'M' pattern are of length $dt/\sqrt{2}$ m.

Upon depicting the trajectory in the pattern of 'M' the node further traverses along the horizontal direction for the same length dt m. The anchor node continues to traverse in the series of the pattern 'M' followed by a horizontal line until it reaches the end of the network area. On reaching the end of the network area, the node traverses vertically with the distance of $3dt$ m in upward direction. Further the anchor node moves horizontally towards the left direction and stops at the position where its coordinate of x is the same as the midpoint of last 'M' traced by it. To overcome the collinearity issue by making a triangle like pattern, a variation with adjacent rows is created by this movement to the x coordinate of the midpoint of the last 'M'. Then, the anchor node traverses in the pattern of alphabet 'W' and a horizontal line of dt m in the reverse direction. The pattern continues till the anchor node reaches the end of network area. Upon reaching the end, the anchor node will again travel in vertical direction for dt m. This pattern of 'M' with the horizontal line towards right direction and 'W' with a horizontal line towards the anchor node is repeated so as to cover the network area.

8.3.3 Non-collinear messages

The mobile anchor node transmits the beacon signal to the other nodes within its transmission range. Once an unknown node receives three different non-collinear beacon messages, then the node can estimate its position with this information. Figure 8.5 shows the triangle-like shape formation between the anchor points which assures that all the nodes deployed in the network are assured to get messages from three non-collinear anchor points. For example,

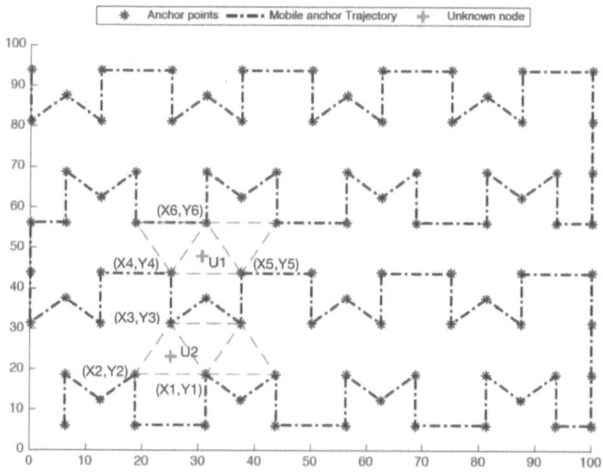

FIGURE 8.5: Proposed path in M-Curves

consider the unknown nodes U1 and U2 in the network will get three anchor points with coordinates (X1, Y1), (X2, Y2), (X3, Y3) and (X4, Y4), (X5, Y5), and (X6, Y6), respectively. When an unknown node does not receive at least three different non-collinear beacon messages it will wait till it receives the messages. Let three non-collinear messages received by unknown node be represented by the matrix,

$$MSG = \begin{bmatrix} X3 - X2 & Y3 - Y2 \\ X2 - X1 & Y2 - Y1 \end{bmatrix}$$

To check the collinearity of the received beacon messages, determinant of the matrix MSG is determined. The received points are non-collinear iff

$$|MSG| = (X3 - X2)(Y2 - Y1) - (Y3 - Y2)(X2 - X1) \neq 0$$

Hence, this assures that the three anchor points chosen for the localization are non-collinear.

8.3.4   Node localization process

The main aim of the node localization process in WSN is to calculate the coordinates of the unknown sensor nodes by minimizing the objective function. The localization process is solved by various optimization techniques by modelling it as an optimization problem. The process of localization using DSA is explained as follows.

1. Initialize a set of m target nodes; a mobile anchor node is sent in the proposed trajectory for localizing the unknown nodes.

2. An additive Gaussian noise is added with the distance between anchor node and unknown node for getting noisy range measurements on the resemblance of real time scenario. Each unknown node calculates the distance using $\widehat{d_i} = d_i + n_i$, where d_i is the actual distance which is calculated between the unknown node's coordinates (x,y) and the anchor points' coordinates (X_i, Y_i) using the equation

$$\sqrt{((x - X_i)^2 + (y - Y_i)^2)} \leq d_i$$

3. The variable n_i is the noise added to the actual distance distributed in range $d_i \pm di\left(\frac{Pn}{100}\right)$, where Pn is the noise percentage in the actual distance.

4. The unknown node is said to be localizable if it receives a beacon signal from at least three anchor nodes.

5. DSA is run independently, for all the unknown nodes which is localizable, to estimate the coordinates. The dolphins are initialized around

the centroid of the anchor points which are in transmission range of the unknown node. Using the following equation, the centroid can be computed.

$$(x_c, y_c) = \left(\frac{1}{N} \sum_{i=1}^{n} x_i, \frac{1}{N} \sum_{i=1}^{n} y_i \right) \tag{8.25}$$

where n denotes the number of anchor points within the transmission range of localizable node.

6. DSA algorithm is used to estimate the coordinates of the unknown node by minimizing the localization error. The objective function of node localization process is the mean square distance between the anchor node and the unknown node which is formulated as

$$Fitness(x, y) = \frac{1}{N} \left(\sum_{i=1}^{N} \sqrt{((X - x_i)^2 + (Y - y_i)^2)} - d_i^2 \right) \tag{8.26}$$

where $N \geq 3$ is the number of anchor points within the transmission range of the unknown node.

7. The unknown node estimates its location by minimizing the objective function after running the algorithm for a number of generations.

This entire node localization process is shown in Figure 8.2.

8.4 Results and discussion

8.4.1 Performance setup

To assess the performance of our proposed model, we considered four path planning models for mobile anchor based localization. The static path planning models used are SCAN, HILBERT, LMAT and Z-curves.

8.4.2 Parameter setup

The proposed model and the other models considered are implemented in MATLAB environment. The parameters for the implementation are considered with respect to the existing literature for easier comparative study of the results. The network field where the unknown sensor nodes are deployed is assumed to be a square area as mentioned before, of size 100×100 m^2. A set S of 200 unknown sensor nodes and a mobile anchor node MA is used for simulation. The transmission range of the mobile anchor node varies with respect to resolution (dt). The transmission range of the anchor node is set to $tr = \sqrt{\frac{5}{2}}d$. Resolution(dt) is dependent on the level.

8.4.3 Performance analysis

Localization accuracy is a significant metric to assess the performance of any localization algorithm. Accuracy of a model can be determined based on the localization error. Two ways of calculating localization error are average localization error and standard deviation of the localization error. For i^{th} unknown sensor node, the localization error can be calculated as

$$LocError_i = \sqrt{((x_i - estx_i)^2 + (y_i - esty_i)^2)} \qquad (8.27)$$

where (x_i, y_i) denotes the actual coordinates of the node and $(estx_i, esty_i)$ denotes the estimated coordinates of the node. Therfore, the average localization error of all unknown nodes in the network can be calculated using

$$AvgLocError = \frac{1}{N}\left(\sum_{i=1}^{N} LocError_i\right) \qquad (8.28)$$

where N is the number of unknown nodes in the network.

Localization error of various strategies such as SCAN, HILBERT, LMAT, Z-Curves and M-curves is compared in the figures 8.7, 8.8 and 8.9. In SCAN, HILBERT and Z-curves, unknown node's position is estimated by centroid method after receiving three different beacon messages. In our model, we estimated the position by Dolphin Swarm Algorithm using a novel fitness function.

We first ran the simulation with 200 unknown nodes with standard deviation 2, level 2, resolution of 12.5 m for 50 simulation runs.The depicted scenario and the node localization in M-Curves is shown in Figure 8.6. Figure 8.7

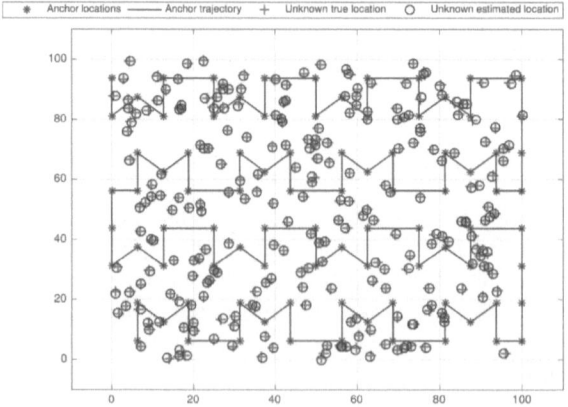

FIGURE 8.6: Localization scenario with M-Curves

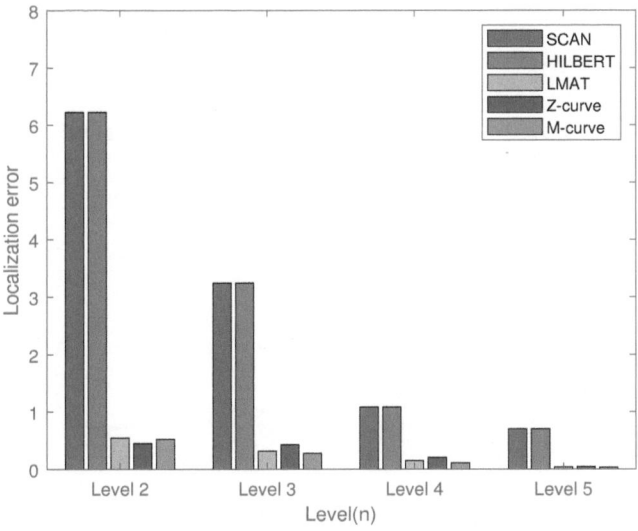

FIGURE 8.7: Localization error vs level with standard deviation of noise $\sigma = 2$

shows the simulation results of localization error versus level. M-Curves trajectory produced the average localization error after 50 runs is 0.523 m. The decrease in localization error can be observed with the increase in level and decrease in resolution. Localization error of M-Curves obtained for level 3 is 0.278 m. With minumum resolution, localization error produced by M-Curves is 0.0328 m. For level 2, SCAN, HILBERT, LMAT and Z-Curves methods produced localization error of 6.22 m, 6.22 m, 0.545 m and 0.4512 m, respectively. Z-Curves gave localization error of 0.038 m which is greater than M-Curves.

Figure 8.8 represents the comparison of localization error for standard deviation 4. For standard deviation 4, M-Curves still performs better than the other trajectories. When $\sigma = 4$, the error for M-curves is 1.336 m and error for SCAN, HILBERT, LMAT and Z-Curves are 7.35, 7.35, 0.7825 and 0.592 m, respectively. It is noteworthy to mention the observation that, for all strategies, increase in level causes decrease in localization error because all the nodes will receive more than three beacon messages so that they can localize themselves. SCAN, HILBERT strategies have the collinearity issue. When the resolution is high, unknown nodes will receive collinear beacon messages with which the unknown nodes estimate their position. The localization error increases if a node uses collinear beacon positions for localization. M-Curves is designed in such a way that each known node receives at least three non-collinear messages. Also, M-Curves and Z-Curves check the collinearity of the beacon nodes before localization. Hence these two models perform better than all other trajectories in simulation. Though, Z-Curves has lesser localization

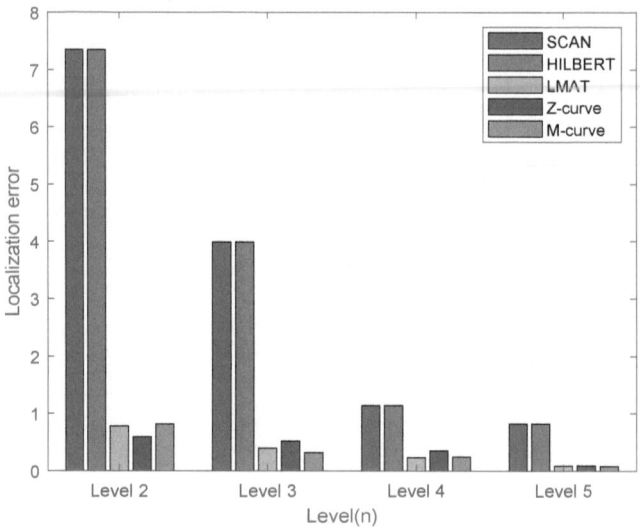

FIGURE 8.8: Localization error vs level with standard deviation of noise $\sigma = 4$

error compared to other models for level 2, the localization error increases with the increase in the level.

The localization error of various strategies when standard deviation is 8 is shown in Figure 8.9. As the noise increases it affects the localization process thereby increasing the localization error. With the increase in level, the error

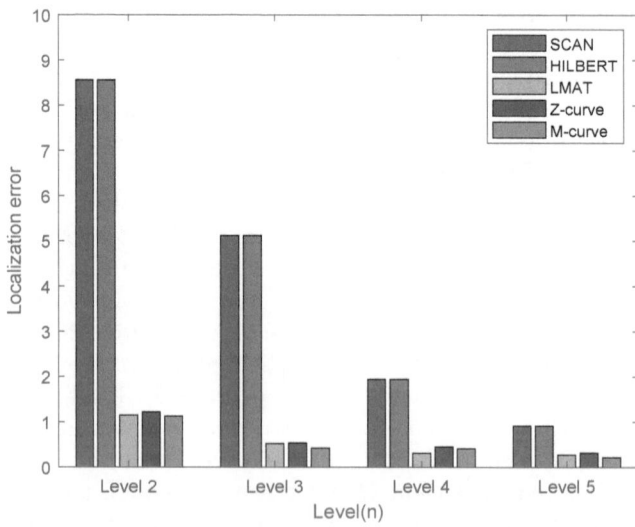

FIGURE 8.9: Localization error vs level with standard deviation of noise $\sigma = 8$

can be reduced but it results in increased path length. The localization error for the M-Curves strategy is 1.126 m which is higher than error produced with less standard deviation. The localization error of SCAN, HILBERT, LMAT and Z-Curves is 8.56 m, 8.56 m, 1.15 m and 1.22 m, respectively. Comparing the results, it is evident that M-Curves performs well in all the cases in terms of localization error.

Standard deviation of the localization error is the measure of deviation of the error from the mean of localization error. High standard deviation of localization error indicates that large portions of the error are far from the mean. A model with high standard deviation of localization error is not considered to be a good model as the localization error varies abnormally. The standard deviation of the localization error is given by

$$LocError_{std} = \sqrt{\frac{1}{N} \sum_{i=1}^{n} (LocError_i - LocError_{mean})^2} \tag{8.29}$$

where N is the number of unknown nodes deployed in the network, $LocError_i$ denotes the localization error of ith node and $LocError_{mean}$ denotes the mean localization error.

We computed the mean of standard deviation values for 50 simulation runs for different levels. Figure 8.10 shows the standard deviation of localization error for different levels with standard deviation of noise $\sigma = 2$. All models show less standard deviation of error, implying that localization errors were close to mean. When level increases the standard deviation of LMAT and M-Curves decreases and has the error values closer to mean.

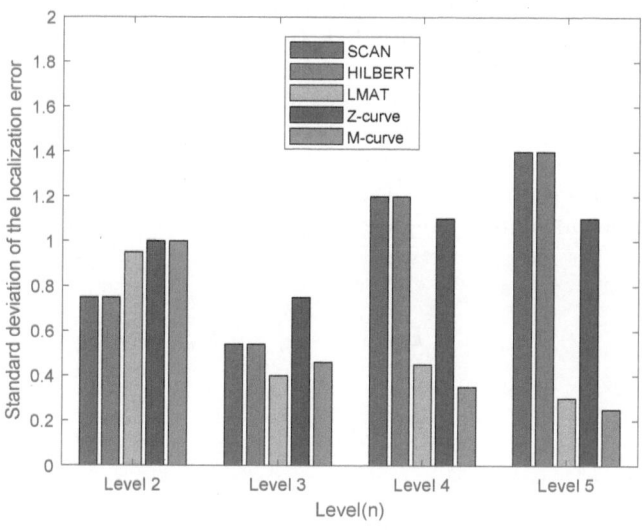

FIGURE 8.10: Level vs standard deviation of error

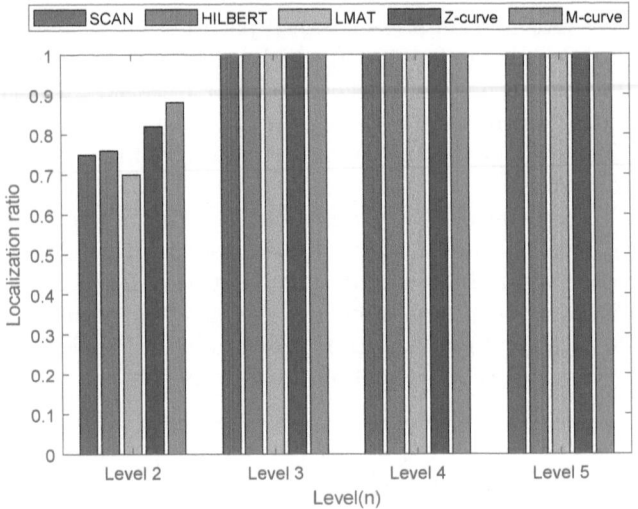

FIGURE 8.11: Localization ratio vs level

Localization ratio is the ratio of number of nodes localized to the number of unknown nodes deployed in the region. Number of localized nodes will be high for any desired model. We considered a scenario with 200 unknown nodes and different levels with standard deviation 2. Localization ratio for various model is shown in Figure 8.11. For level 2, the localization ratio of SCAN and HILBERT are 0.75 m and 0.76 m, respectively, while LMAT, Z-Curves and M-Curves have 0.7 m, 0.82 m and 0.88 m, respectively. With increase in level, all the models achieved full coverage. That is, all the nodes are localized in the network although they give different results in terms of other factors.

Path length is the magnitude of the distance covered by the mobile anchor to follow the proposed model in the network. Though the path length did not have any effect in localization error, it helps in finding the time required for the entire localization process. Path length of SCAN, HILBERT, LMAT, Z-Curves can be calculated using the following equations, where dt_m represents the distance (in meters) between two consecutive points in the trajectory [38, 40, 39].

$$PL_{SCAN} = \frac{L^2}{dt_m} - dt_m \tag{8.30}$$

$$PL_{HILBERT} = \frac{L^2}{dt_m} \tag{8.31}$$

$$PL_{LMAT} = \frac{L^2}{dt_m} + \frac{L}{5} \times dt_m \tag{8.32}$$

$$PL_{Z-Curves} = \left[\left(\frac{5}{8} \times 4^3\right) - 1\right]dt_m + \left[\left(\frac{3}{8} \times 4^3\right)\right] \times \sqrt{2}dt_m \qquad (8.33)$$

$$PL_{M-Curves} = \left[16 \times \left[\left(2 + \sqrt{2}\right)dt_m + dt_m\right]\right] - \frac{dt_m}{2} + 7dt_m \qquad (8.34)$$

Eq. 8.34 shows the path length of our proposed trajectory M-Curves. The two vertical segments of M pattern cover a distance of dt_m. The entire network area which we have considered has 16 such M patterns thus leading to the 16 dt_m term. Similarly the diagonal segments of M cover a distance of $\sqrt{2}dt_m$ thus summing up to a total of $16*\sqrt{2}dt_m$ for the entire network area. Similarly after traversing the M pattern, the anchor node moves horizontally for dt_m distance. Only one horizontal traversal at the initial position of the anchor node has a length of $0.5 * dt_m$ thus leading to a total of $15.5dt_m$ which can be generally written as number of $M's \times (dt_m - 0.5dt_m)$. The term $7dt_m$ represents the distance taken by the movement of the path from one row to another.

Path length of the various strategies is shown in the Figure 8.12. The distance travelled by SCAN, HILBERT, LMAT, Z-Curves and M-Curves is 787.5 m, 800 m, 1050 m, 912 m and 964.1 m, respectively. LMAT and M-Curves have path length greater than other models. M-Curves gives better localization accuracy compared to LMAT with lesser path length.

To evaluate the performance of DSA optimized localization process, APT technique is used. We ran the proposed path model with both the techniques. The comparison results are shown in the Figure 8.13. DSA optimized localization process gives low localization error compared with APT in all levels as shown in the Figure 8.13. From the results it is evident that optimization reduces the localization error which in turn increases the localization accuracy.

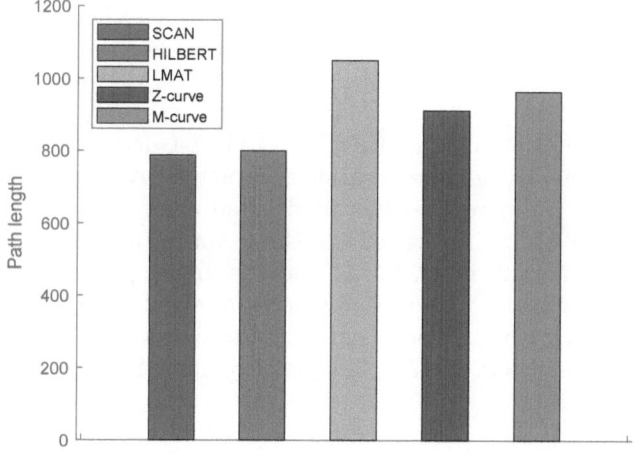

FIGURE 8.12: Different static models and its path length

Wireless Sensor Networks

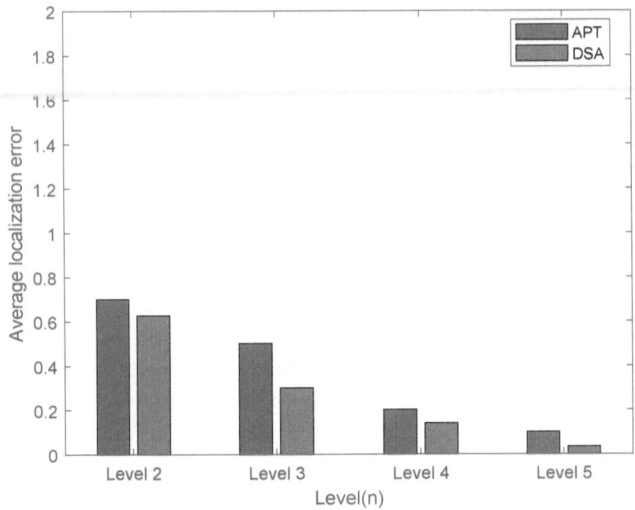

FIGURE 8.13: Comparision of APT and DSA on the proposed model

8.5   Conclusion

We proposed a mobile anchor trajectory named 'M-Curves' for the localization approach based on mobile anchors with the consideration of major features for localization. Also, DSA optimization technique is adopted for localization by optimizing the fitness function. The crucial factors such as full coverage, high localization accuracy and localization ratio are assured by the proposed model. The trajectory guarantees that each unknown node receives at least three non-collinear beacon messages from the anchor node for efficient localization. Localization error is modeled as fitness function and is optimized using DSA. Compared to the traditional techniques, the performance of the optimized localization process is enhanced. Our model outperforms the other static models such as SCAN, HILBERT, LMAT, Z-Curves in terms of metrics such as localization ratio and localization accuracy. SCAN and HILBERT have lesser path length compared to other models but these models are affected by collinearity issues which in turn affects the localization accuracy. In spite of resolving the collinearity issue, LMAT has the higher path length which will cause high consumption of energy. Z-Curves resolved both collinearity issue and path length. However, experimental results shows that M-Curves outperforms SCAN, HILBERT, LMAT, Z-Curves models in terms of localization accuracy.

Chapter 9

Conclusion and Future Research

9.1 Conclusion

In this book, we have presented various evolutionary and swarm intelligence algorithms to defeat different challenges in WSN. We have also addressed various drawbacks associated with existing algorithms and also worked on them to get a superior performance through these optimization algorithms. We have also shown the experimental results of some of the existing works and the proposed algorithms outperform the compared algorithms.

The introductory part of the book has been presented in Chapter 1. It comprises of the overview of the book, an introduction to wireless sensor networks, motivation behind the book, resources used and organization of the chapter. In Chapter 2, the extensive review of the existing load balancing techniques, energy efficient algorithms and localization-based approaches are discussed. Our contribution begins with Chapter 3 and is concluded as follows:

- In Chapter 3, the SCE approach is applied for load balancing of gateways in WSN. A novel fitness function is designed to measure the quality of the solution. The fitness function is designed according to the heavy and underloaded gateways. A new approach is followed for generation of offspring, that only the worst solution is replaced by the better offspring. This takes over the best solutions in next generation. The performance of the proposed approach is compared with state-of-the-art load balancing algorithms. It is observed that the proposed algorithm outperforms these algorithms.

- In Chapter 4, the various improvements in Shuffled Complex Evolution algorithm are applied in order to balance the load of gateways in WSNs. ISCE varies from SCE in terms of phases in initial population generation and generation of offspring phase. A novel phase has been added after offspring generation phase, called relocation phase. The relocation phase helps in minimizing the energy consumption of sensor nodes. Along with this modification, a novel fitness function is designed for measuring quality of solution in terms of load ratio, distance ratio and lifetime of gateways. The ISCE algorithm has been compared with existing load balancing algorithms considering the parameters such as heavy loaded sensor nodes, energy consumption, load balancing for both equal and unequal load, network lifetime parameters. It is observed that our ISCE algorithm outperforms in the above mentioned parameters.

- In Chapter 5, SFLA is improved according to the application of WSN. The improvements have been made in the phases of initial population, and offspring generation. A novel fitness function is designed by considering residual energy of gateways. To prove the effectiveness of proposed approach, it is compared with the state-of-the-art algorithms. It is observed that proposed ISFLA outperforms these algorithms under various evaluation factors.

- In Chapter 6, a clustering approach based on SCE-PSO is proposed for efficient load balancing of gateways in WSNs. A novel fitness feature is also proposed by considering the distance between sensor nodes and gateways. The recommended load balancing algorithm has been compared with state-of-the-art SGA, NGA, NLDLB and SBLB algorithms. The proposed load balancing algorithm was contrasted with state-of-the-art SGA, NGA, NLDLB and SBLB algorithms. Experimental results show that the proposed solution has outperformed in terms of network lifetime, total energy utilization, the minimum lifetime gateway that dissolves its energy and half the gateway die as compared to state-of-the-art algorithms.

- In Chapter 7, the PSO based routing algorithm is used for efficient routing in WSNs to prolong the lifetime of WSNs. For a suitable path, a novel fitness is also designed to take into account the distance, hop-count and relay load factor of the network. The proposed PSO based routing algorithm is compared to state-of-the-art routing approaches such as GA, PSO and GLBCA to analyze the network lifetime of the WSNs. The efficiency of the proposed algorithm is evaluated in terms of network lifetime, number of hops and average relay load under two separate WSN typologies. The experimental results show that the proposed PSO based routing algorithm increased the network lifetime and reduced the average relay load at the cost slightly more number of hops when compared with state-of-the-art routing techniques.

- In Chapter 8, a mobile anchor trajectory named 'M-Curves' for the localization approach based on mobile anchors with the consideration of major features for localization is proposed. Also, DSA optimization technique is adopted for localization by optimizing the fitness function. The crucial factors such as full coverage, high localization accuracy and localization ratio are assured by the proposed model. The trajectory guarantees that each unknown node receives at least three non-collinear beacon messages from the anchor node for efficient localization. Localization error is modeled as fitness function and is optimized using DSA. Compared to the traditional techniques, the performance of the optimized localization process is enhanced. Our model outperforms the other static models such as SCAN, HILBERT, LMAT, Z-Curves in terms of metrics such as localization ratio and localization accuracy. SCAN and HILBERT have a lesser path length compared to other models but these models are affected by a collinearity issue which in turn affects the localization accuracy. In spite of resolving the collinearity issue, LMAT has the higher path length which will cause high consumption of energy. Z-Curves resolved both collinearity issue and path length. However, experimental results show that M-Curves outperforms SCAN, HILBERT, LMAT, Z-Curves models in terms of localization accuracy.

In summary, the book presents the six bio-inspired approaches to defeat various challenges in WSN, such as load balancing, energy efficiency, clustering, routing and localization. It shows that all the bio-inspired approaches for all these challenges outperform the compared heuristic approaches under various evaluation factors.

9.2   Future research

Although the proposed approaches have been found to be efficient in improving the performance under defined evaluation factors, the following future research is proposed for further improvements.

- A novel bio-inspired algorithm can be designed to improve the complexity of the problems.

- Various soft computing techniques can be applied along with the bio-inspired approaches.

- The proposed approaches can be applied to other challenges in WSN, such as data aggregation, data collection and communication, coverage and connectivity, etc.

- These approaches can be applied to the mobile networks, in which mobile sensor and a sink or multiple sinks can be adoptable.

- There is a scope for selection of the solutions by estimating the energy consumption of each sensor node at early stages, so that the lifetime of the network can be prolonged.

Bibliography

[1] Ian F. Akyildiz, Weilian Su, Yogesh Sankarasubramaniam, and Erdal Cayirci. A survey on sensor networks. *IEEE Communications Magazine*, 40(8):102-114, 2002.

[2] Ravinder Kaur and Kamal Preet Singh. An efficient multipath dynamic routing protocol for mobile WSNs. *Procedia Computer Science*, 46:1032-1040, 2015.

[3] Palvinder Singh Mann, Satvir Singh, and Anil Kumar. Computational intelligence based meta heuristic for energy-efficient routing in wireless sensor networks. In *Congress on Evolutionary Computation (CEC)*, pages 4460-4467. IEEE, 2016.

[4] Amrit Pal Singh, Parminder Singh, and Rakesh Kumar. A review on impact of sinkhole attack in Wireless Sensor Networks. *International Journal of Advanced Research in Computer Science and Software Engineering*, 5(8), August 2015.

[5] P. V. Mane Deshmukh, S. C. Pathan, S. V. Chanvan, S. K. Tilekar, and B. P. Ladgaonkar. Wireless Sensor Network for monitoring of air pollution near industrial sector. *International Journal of Advanced Research in Computer Science and Software Engineering* 6(6), June 2016.

[6] Y. Xu, J. Heidemann, and D. Estrin. Geography-informed energy conservation for ad hoc routing, *Procidia Mobicom*, pages 70-84, 2001.

[7] Kumar Vipin and Kumar Sushil. Energy balanced position-based routing for lifetime maximization of wireless sensor networks, *Ad Hoc Networks*, 2016.

[8] A. Giuseppe. Energy conservation in wireless sensor networks: a survey, *Ad Hoc Networks*, 7:537-568, 2009.

[9] C.Y. Chong and S.P. Kumar. Sensor networks: evolution, opportunities, and challenges, *Proceedings of the IEEE*, 91(8):1247-1256, 2003.

[10] L. Emanuele. Energetic sustainability of routing algorithms for energy harvesting wireless sensor networks, *Computer Communication*, 30:2976-2986, 2007.

[11] M. J. Handy, Haase Marc, and Timmermann Dirk. Low energy adaptive clustering hierarchy with deterministic cluster-head selection. In *4th international workshop on mobile and wireless communications network*, Stockholm, Sweden, pages 368-372. IEEE, 2002.

[12] Naveen Kumar and Jasbir Kaur. Improved leach protocol for wireless sensor networks. In *7th International Conference on Wireless Communications, Networking and Mobile Computing*, Wuhan, China, pages 1-5. IEEE, 2011.

[13] Pratyay Kuila and Prasanta K. Jana. Energy efficient load-balanced clustering algorithm for wireless sensor networks. *Procedia Technology*, 6:771-777, 2012.

[14] Jing Zhang and Ting Yang. Clustering model based on node local density load balancing of wireless sensor networks. *Fourth international conference on emerging intelligent data and web technologies*, 2013.

[15] Vaishali S. Gattani and S.M. Haider Jafri. Data collection using score based load balancing algorithm in wireless sensor networks. In *International Conference on Computing Technologies and Intelligent Data Engineering (ICCTIDE)*, IEEE, 2016.

[16] B. Baranidharan and B. Santhi. DUCF: Distributed load balancing unequal clustering in wireless sensor networks using Fuzzy approach, *Applied Soft Computing*, 40:495-506, 2016.

[17] P. Chanak, I. Banerjee, and H. Rahaman. Load management scheme for energy holes reduction in wireless sensor networks. *Computers and Electrical Engineering*, Elsevier, 48:343-357, 2015.

[18] Salehi Panahi, S. Morteza, and Mortaza Abbaszadeh. Proposing a method to solve energy hole problem in wireless sensor networks. *Alexandria Engineering Journal*, Elsevier, 2017.

[19] P. Kuila, S. K. Gupta, and P. K. Jana. A novel evolutionary approach for load balanced clustering problem for wireless sensor networks. *Swarm and Evolutionary Computation*. Elsevier, 12:48-56, 2013.

[20] N. A. Al-Aboody and H. S. Al-Raweshidy. Grey wolf optimization-based energy-efficient routing protocol for heterogeneous wireless sensor networks. In *Fourth IEEE International Symposium on Computational and Business Intelligence (ISCBI)*, Olten, Switzerland, pages 101-107, 2016 Sep 5.

[21] J. W. Lee and J. J. Lee. Ant-Colony-Based Scheduling Algorithm for Energy-Efficient Coverage of WSN, *IEEE Sensors Journal*, 12(10):3036-3046, Jul. 2012.

[22] Y. Lin, J. Zhang, H.S.H. Chung, W.H. Ip, Y. Li, and Y.H. Shi. An ant colony optimization approach for maximizing the lifetime of heterogeneous wireless sensor networks. *IEEE Transactions on Systems, Man, and Cybernetics, Part C (Applications and Reviews)*, 42(3):408-420, 2011.

[23] R. W. Gan, Q. S. Guo, and H. Y. Chang. Improved ant colony optimization algorithm for the traveling salesman problems, *Journal of System Engineering and Electronics*, 21(2):329-333, 2010.

[24] S. Mehrjoo, H. Aghaee, and H. Karimi. A novel hybrid ga-abc based energy efficient clustering in wireless sensor network. *Canadian Journal on Multimedia and Wireless Network*, 2(2), 2011.

[25] S. Selvakennedy, S. Sinnappan, and Y. Shang. T-ant: a nature-inspired data gathering protocol for wireless sensor networks. *Journal of Communications*, 1(2):22-29, 2006.

[26] N. M. A. Latiff, C. C. Tsimenidis, and B. S. Sharif. Energy-aware clustering for wireless sensor networks using particle swarm optimization. In *18th International Symposium on Personal, Indoor and Mobile Radio Communications*, pages 1-5, 2007.

[27] W. B. Heinzelman, A. P. Chandrakasan, H. Balakrishnan, and C. MIT. An application-specific protocol architecture for wireless microsensor networks. *IEEE Transactions on Wireless Communications*, 1(4):660-670, 2002.

[28] I. F. Akyildiz, W. Su, Y. Sankarasubramaniam, and E. Cayirci. Wireless sensor networks: a survey, *Computer Networks*, 38(4), pages 393-422, 2002.

[29] X. Li, N. Mitton, I. Simplot-Ryl, and D. Simplot-Ryl. Dynamic beacon mobility scheduling for sensor localization, *IEEE Transactions on Parallel and Distributed Systems*, 23(8):1439-1452, 2012.

[30] M. Moradi, J. Rezazadeh, and A. S. Ismail. A reverse localization scheme for underwater acoustic sensor networks, *IEEE Sensors Journal*, 12(4):4352-4380, 2012.

[31] S. Halder and A. Ghosal. A survey on mobility-assisted localization techniques in wireless sensor networks, *Journal of Network and Computer Applications*, 60:82-94, 2016.

[32] L. Chelouah, F. Semchedine, and L. Bouallouche-Medjkoune. Localization protocols for mobile wireless sensor networks: A survey, *Computers & Electrical Engineering*, 2017.

[33] Mihail L. Sichitiu and Vaidyanathan Ramadurai. Localization of wireless sensor networks with a mobile beacon. In *IEEE International Conference on Mobile Ad-hoc and sensor systems*, pages 174-183. IEEE, 2004.

[34] K.-F. Ssu, C.-H. Ou, and H. C. Jiau. Localization with mobile anchor points in wireless sensor networks, *IEEE Transactions on Vehicular Technology*, 54(3):1187-1197, 2005.

[35] S. Lee, E. Kim, C. Kim, and K. Kim. Localization with a mobile beacon based on geometric constraints in wireless sensor networks. *IEEE Transactions on Wireless Communications*, 8(12):5801-5805, 2009.

[36] A. Savvides, C.-C. Han, and M. B. Strivastava. Dynamic fine-grained localization in ad-hoc networks of sensors, In *Proceedings of the 7th annual international conference on Mobile computing and networking.* ACM, pages 166-179. 2001.

[37] N. Bulusu, J. Heidemann, and D. Estrin. GPS-less low-cost outdoor localization for very small devices, *IEEE Personal Communications*, 7(5):28-34, 2000.

[38] D. Koutsonikolas, S. M. Das, and Y. C. Hu. Path planning of mobile landmarks for localization in wireless sensor networks, *Computer Communications*, 30(13):2577-2592, 2007.

[39] J. Rezazadeh, M. Moradi, A. S. Ismail, and E. Dutkiewicz. Superior path planning mechanism for mobile beacon-assisted localization in wireless sensor networks, *IEEE Sensors Journal*, 14(9):3052-3064, 2014.

[40] J. Jiang, G. Han, H. Xu, L. Shu, and M. Guizani. LMAT: Localization with a mobile anchor node based on trilateration in wireless sensor networks, In *Global Telecommunications Conference (GLOBECOM 2011)*, IEEE, pages 1-6. 2011.

[41] Abdullah Alomari, Frank Comeau, William Phillips, and Nauman Aslam. New path planning model for mobile anchor-assisted localization in wireless sensor networks. *Wireless Networks*, 24(7):2589-2607. 2018.

[42] U. Nazir, N. Shahid, M. Arshad, and S. H. Raza. Classification of localization algorithms for wireless sensor network: A survey, In *International conference on Open source systems and technologies (ICOSST)*, IEEE, pages 1-5. 2012.

[43] J. Blumenthal, R. Grossmann, F. Golatowski, and D. Timmermann. Weighted centroid localization in ZigBee-based sensor networks. In *International Symposium on Intelligent Signal Processing*, IEEE, pages 1-6. 2007.

[44] Q. Dong and X. Xu. A novel weighted centroid localization algorithm based on RSSI for an outdoor environment, *Journal of Communications*, 9(3):279-285, 2014.

[45] T. He, C. Huang, B. M. Blum, J. A. Stankovic, and T. Abdelzaher. Range-free localization schemes for large scale sensor networks, In *Proceedings of the 9th annual international conference on Mobile computing and networking*. ACM, pages 81-95. 2003.

[46] R. C. Luo, O. Chen, and S. H. Pan. Mobile user localization in wireless sensor network using grey prediction method, In 31st *Annual Conference of Industrial Electronics Society (IECON)*, IEEE, pages 6-12. 2005.

[47] J.-P. Sheu, W.-K. Hu, and J.-C. Lin. Distributed localization scheme for mobile sensor networks, *IEEE Transactions on Mobile Computing*, 9(4):516-526, 2010.

[48] B. Neuwinger, U. Witkowski, and U. Ruckert. Ad-hoc communication and localization system for mobile robots, In *FIRA RoboWorld Congress*. Springer, pages 220-229, 2009.

[49] W. Wang and Q. Zhu. Sequential Monte Carlo localization in mobile sensor networks. *Wireless Networks*, 15(4):481-495, 2009.

[50] T.-q. Wu, M. Yao, and J.-h. Yang. Dolphin swarm algorithm, *Frontiers of Information Technology & Electronic Engineering*, 17(8):717-729, 2016.

[51] A. Gopakumar and L. Jacob. Localization in wireless sensor networks using particle swarm optimization, *IET Conference on Wireless, Mobile and Multimedia Networks*, 2008.

[52] R. V. Kulkarni and G. K. Venayagamoorthy. Particle swarm optimization in wireless sensor networks: A brief survey, *IEEE Transactions on Systems, Man, and Cybernetics, Part C (Applications and Reviews)*, 41(2):262-267, 2011.

[53] R. V. Kulkarni, G. K. Venayagamoorthy, and M. X. Cheng. Bio-inspired node localization in wireless sensor networks, In *International Conference on Systems, Man and Cybernetics*, IEEE, pages 205-210, 2009.

[54] Z. Yang and Y. Liu. Quality of trilateration: Confidence-based iterative localization, *IEEE Transactions on Parallel and Distributed Systems*, 21(5):631-640, 2010.

[55] S. Okdem. A real-time noise resilient data link layer mechanism for unslotted IEEE 802.15. 4 networks, *International Journal of Communication Systems*, 30(3), 2017.

[56] J. J. Perez-Solano, J. M. Claver, and S. Ezpeleta. Optimizing the MAC protocol in localization systems based on IEEE 802.15. 4 networks, *IEEE Sensors*, 17(7), 2017.

[57] M. Rengasamy, E. Dutkiewicz, and M. Hedley. MAC design and analysis for wireless sensor networks with cooperative localisation. In *International Symposium on Communications and Information Technologies*, IEEE, pages 942-947. 2007.

[58] Q. Y. Duan, Vijai K. Gupta, and Soroosh Sorooshian. Shuffled complex evolution approach for effective and efficient global minimization, *Journal of Optimization Theory and Applications*, 76(3):501-521, 1993.

[59] Kasthurirangan Gopalakrishnan and Sunghwan Kim. Global optimization of pavement structural parameters during back-calculation using hybrid shuffled complex evolution algorithm. *Journal of Computing in Civil Engineering*, 24(5):441-451, 2010.

[60] Fuqing Zhao. An improved shuffled complex evolution algorithm with sequence mapping mechanism for job shop scheduling problems, *Expert Systems with Applications*, 42(8):3953-3966, 2015.

[61] Kasthurirangan Gopalakrishnan and Sunghwan Kim. Global optimization of pavement structural parameters during back-calculation using hybrid shuffled complex evolution algorithm. *Journal of Computing in Civil Engineering*, 24(5):441-451, 2010.

[62] Anju Bala and Aman Kumar Sharma. A comparative study of modified crossover operators. In *Third International Conference on Image Information Processing (ICIIP)*, IEEE, 2015.

[63] Ataul Bari. A genetic algorithm based approach for energy efficient routing in two-tiered sensor networks, *Ad Hoc Networks*, 7(4):665-676, 2009.

[64] Muzaffar Eusuff, Kevin Lansey and Fayzul Pasha. Shuffled frog-leaping algorithm: a memetic meta-heuristic for discrete optimization, *Engineering optimization*, 38(2):129-154, 2006.

[65] G. G. Samuel and C. Christober Asir Rajan. A Modified Shuffled Frog Leaping Algorithm for Long-Term Generation Maintenance Scheduling. In *Proceedings of the Third International Conference on Soft Computing for Problem Solving*. Springer, New Delhi, 2014.

[66] Waltenegus Dargie and Christian Poellabauer. Fundamentals of Wireless Sensor Networks: Theory and Practice. 2011.

[67] Jun Zheng and Abbas Jamalipour. Wireless sensor networks: a networking perspective. John Wiley & Sons, 2009.

[68] El Emary, M. M. Ibrahiem, and S. Ramakrishnan. Wireless sensor networks: from theory to applications. CRC Press, 2013.

[69] Edgar H. Callaway, Jr. Wireless sensor networks: architectures and protocols. CRC Press, 2003.

[70] Robert Faludi. Building wireless sensor networks: with ZigBee, XBee, arduino, and processing. O'Reilly Media, Inc., 2010.

[71] Mustapha Reda Senouci and Abdelhamid Mellouk. Deploying wireless sensor networks: theory and practice. Elsevier, 2016.

[72] Alexey N. Averkin, A. G. Belenki, and G. Zubkov. Soft Computing in Wireless Sensors Networks. EUSFLAT Conf.(1). 2007.

[73] Yingshu Li and My T. Thai, eds. Wireless sensor networks and applications. Springer Science & Business Media, 2008.

[74] Chee Peng Lim and Satchidananda Dehuri, eds. Innovations in swarm intelligence. Vol. 248. Springer Science & Business Media, 2009.

[75] Xin-She Yang, Zhihua Cui, Renbin Xiao, Amir Hossein Gandomi, and Mehmet Karamanoglu, eds. Swarm intelligence and bio-inspired computation: theory and applications. Newnes, 2013.

[76] Kaisa Miettinen and P. Preface By-Neittaanmaki. Evolutionary algorithms in engineering and computer science: recent advances in genetic algorithms, evolution strategies, evolutionary programming, GE. John Wiley & Sons, Inc., 1999.

[77] Ajith Abraham, Crina Grosan, and Hisao Ishibuchi, eds. Hybrid Evolutionary Algorithms. Springer-Verlag Berlin Heidelberg, 2007.

[78] Kalyanmoy Deb. Multi-objective optimization using evolutionary algorithms. Vol. 16. John Wiley & Sons, 2001.

[79] Dipankar Dasgupta and Zbigniew Michalewicz, eds. Evolutionary algorithms in engineering applications. Springer Science & Business Media, 2013.

[80] Nancy Arana-Daniel, Carlos Lopez-Franco, and Alma Y. Alanis. Bio-inspired algorithms for engineering. Butterworth-Heinemann, 2018.

[81] Shangce Gao, ed. Bio-Inspired Computational Algorithms and Their Applications. BoD–Books on Demand, 2012.

[82] Heinz Mühlenbein, M. Schomisch, and Joachim Born. The parallel genetic algorithm as function optimizer. *Parallel Computing*, 17(6-7):619-632, 1991.

[83] Seyedali Mirjalili, Seyed Mohammad Mirjalili, and Andrew Lewis. Grey wolf optimizer. *Advances in Engineering Software*, 69: 46-61, 2014.

[84] M. A. O. Song and Cheng-lin Zhao. Unequal clustering algorithm for WSN based on fuzzy logic and improved ACO. *The Journal of China Universities of Posts and Telecommunications*, 18(6): 89-97, 2011.

[85] Dervis Karaboga and Bahriye Basturk. A powerful and efficient algorithm for numerical function optimization: artificial bee colony (ABC) algorithm. *Journal of Global Optimization*, 39(3):459-471, 2007.

[86] Wendi B. Heinzelman, Anantha P. Chandrakasan, and Hari Balakrishnan. Application specific protocol architecture for wireless microsensor networks, *IEEE Transactions on Wireless Communications*, 1(4):660-670, 2002.

[87] Jiang Yan, Hu Tiesong, Huang Chongchao, Wu Xianing, and Gui Faling. A shuffled complex evolution of particle swarm optimization algorithm, In *International Conference on Adaptive and Natural Computing Algorithms*, pages 341-349, 2007.

[88] Russell Eberhart and James Kennedy. A new optimizer using particle swarm theory. In *Proceedings of the Sixth International Symposium on Micro Machine and Human Science*, pages 39-43, IEEE, 1995.

[89] Riccardo Poli, James Kennedy, and Tim Blackwell. Particle swarm optimization, *Swarm Intelligence*, 1(1):33-57, 2007.

[90] Pratyay Kuila and Prasanta K. Jana. Energy efficient clustering and routing algorithms for wireless sensor networks: Particle swarm optimization approach, *Engineering Applications of Artificial Intelligence*, 33:127-140, 2014.

[91] Chih-Chung Lai, Chuan-Kang Ting, and Ren-Song Ko. An effective genetic algorithm to improve wireless sensor network lifetime for large-scale surveillance applications. *IEEE Congress on Evolutionary Computation*, pages 3531-3538, 2007.

[92] Shuai Yu, Rui Wang, Hongke Xu, Wanggen Wan, Yueyue Gao, and Yanliang Jin. WSN nodes deployment based on artificial fish school algorithm for traffic monitoring system. IET, 2011.

[93] Huan Zhao, Qian Zhang, Liang Zhang, and Yan Wang. A novel sensor deployment approach using fruit fly optimization algorithm in wireless sensor networks. In *Trustcom/BigDataSE/ISPA*, 1:1292-1297. IEEE, 2015.

[94] Wei Hong Lim and Nor Ashidi Mat Isa. Particle swarm optimization with increasing topology connectivity. *Engineering Applications of Artificial Intelligence*, 27:80-102, 2014.

[95] Habib M. Ammari, and Sajal K. Das. A trade-off between energy and delay in data dissemination for wireless sensor networks using transmission range slicing. *Computer Communications*, 31(9):1687-1704, 2008.

[96] Chor Ping Low, Can Fang, Jim Mee Ng, and Yew Hock Ang. Efficient load-balanced clustering algorithms for wireless sensor networks. *Computer Communications*, 31(4):750-759, 2008.

Index